T0133734

Advanced High Dynamic Range Imaging

SECOND EDITION

Advanced High Dynamic Range Imaging

SECOND EDITION

Francesco Banterle
Alessandro Artusi
Kurt Debattista
Alan Chalmers

CRC Press
Taylor & Francis Group
Boca Raton London New York

CRC Press is an imprint of the
Taylor & Francis Group, an **informa** business
AN A K PETERS BOOK

CRC Press
Taylor & Francis Group
6000 Broken Sound Parkway NW, Suite 300
Boca Raton, FL 33487-2742

© 2018 by Taylor & Francis Group, LLC
CRC Press is an imprint of Taylor & Francis Group, an Informa business

No claim to original U.S. Government works

Printed on acid-free paper
Version Date: 20170613

International Standard Book Number-13: 978-1-4987-0694-0 (Hardback)

Library of Congress Cataloging-in-Publication Data

Names: Banterle, Francesco, author. | Artusi, Alessandro, author. |
Debattista, Kurt, 1975- author. | Chalmers, Alan, author.
Title: Advanced high dynamic range imaging / Francesco Banterle, Alessandro
Artusi, Kurt Debattista, Alan Chalmers.
Description: Second edition. | Boca Raton : Taylor & Francis, CRC Press,
2018. | Revised edition of: Advanced high dynamic range imaging /
Francesco Banterle ... [et. al.]. 2011. | Includes bibliographical
references and index.
Identifiers: LCCN 2017012410 | ISBN 9781498706940 (hardback : alk. paper)
Subjects: LCSH: High dynamic range imaging. | Image processing--Digital
techniques. | Photography--Exposure.
Classification: LCC TR594 .B36 2018 | DDC 771/.4--dc23
LC record available at https://lccn.loc.gov/2017012410

Visit the Taylor & Francis Web site at
http://www.taylorandfrancis.com

and the CRC Press Web site at
http://www.crcpress.com

To my brothers. —FB

Dedicated to all of you: Franca, Nella, Sincero, Marco, Giancarlo, and Despo. You are always in my mind. —AA

To Maya. Glad you could join us too! —KD

To Eva, Erika, Andrea, and Thomas. You are my reality!
—AC

Contents

Foreword

High dynamic range (HDR) imaging has become an essential topic across all of the imaging disciplines – including photography, computer graphics, image processing, and computer vision. HDR is used in diverse applications including the creation of fine art, and the collection of data for scientific analysis. The first edition of *Advanced High Dynamic Range Imaging* presented a complete treatment of the subject ranging from the acquisition of HDR to be used as data for lighting computer graphics scenes to the processing of HDR for display on a wide selection of devices. The first edition covered the theory of HDR, and also provided practical MATLAB® code for the reader to use. I used the first edition both in a computer graphics course on "Photography-Based Computer Graphics" and in various student research projects for acquiring light scattering properties of materials.

Over the past 5 years, not only has the usage of HDR increased, but also research in HDR has continued and accelerated. This second edition of the book covers the important recent advances. There are key updates on many of the topics that were already active at the time of the first edition – acquisition, color management, tone mapping, adapting low dynamic range content to HDR displays, and metrics. In these areas, many initial techniques have been replaced by sophisticated and mature approaches.

The most significant recent advance included in this edition, however, is the emergence of high dynamic range video as a practical medium. High dynamic range video required substantial research rather than simply applying previous work for still images to multiple images. For acquisition, affordable hardware that could rapidly take multiple exposures was needed, and effective de-ghosting algorithms were needed to combine the exposures of changing scenes. Tone mapping and color management for video requires temporal consistency, and investigation of new types of potential perceptual artifacts. The sheer quantity of data for video demands more effective compression schemes.

This new edition is a timely update and expansion of the original text. Practitioners, educators, and researchers will all find it useful in their work.

—Holly Rushmeier
Yale University

Preface

The human visual system (HVS) has not changed much in the last 40,000 years. Imaging techniques, on the other hand, have changed dramatically in just the last 5 years, since the first edition of this book appeared. While the HVS is able to see up to 20 f-stops of lighting in a scene with minimal eye adaption, and from starlight through to bright sunshine with full adaption, traditional imaging (or low dynamic range [LDR] aka standard dynamic range [SDR]) techniques are only capable of capturing 8 stops. This is now rapidly changing. As Figure 1 shows, high dynamic range (HDR) is providing a step change in imaging with its capablity of capturing and displaying detail in the very brightest and very darkest parts of a scene simultaneously. Indeed HDR can (theoretically) capture, store, transmit, and deliver even more than the HVS can see at any adaption level.

When the first edition of this book appeared in 2011, HDR video was just beginning to emerge from the research labs to more mainstream use. Although static HDR imaging had been around for more than 20 years, the ability to capture and display moving scenes in HDR was still in its infancy. Major breakthroughs, such as commercial HDR displays from SIM2 in 2009 and, also in 2009, a prototype video camera from the German high-precision camera manufacturer Spheron VR and the University of Warwick, UK, which could capture 20 f-stops (1920 × 1080 resolution) at 30 frames a second and play this directly on SIM2's display, clearly showed the potential step change that HDR could bring about. The subsequent consumer drive for HDR was not long in coming. With 3D televisions struggling to be widely adopted, and the uptake of 4K resolution televisions experiencing sluggish growth, HDR was seized upon by manufacturers as a way of providing new products that would be attractive to consumers as they would be clearly visually "better" than the previous generation [67, 87]. By the Consumer Electronics Show (CES) in January 2016, a number of

Figure 1. Capturing all the detail in a scene with HDR. (Cartoon courtesy of Alan Chalmers and Lance Bell.)

leading manufacturers, including Sony, LG, Panasonic and Samsung, all showed TVs claiming to support HDR, while camera manufacturers, such as ARRI, Grass Valley and RED, were able to announce video cameras capable of capturing > 14 f-stops. To support this drive to consumer HDR, standards bodies are already considering two proposals for HDR: HDR10 [270] and HLG [73]. However, this first wave of consumer HDR has serious limitations and is not able to deliver the full potential of HDR. Currently HDR content has to be viewed in a dark room in order to see all the detail, and available cameras and displays are only capable of capturing and displaying a dynamic range which is well below that which can be seen in a real scene [87]. The aim of this book is to provide a comprehensive practical guide to the very latest developments in HDR. By examining the key problems associated with all facets of HDR imaging and providing a detailed foundation, together with supporting MATLAB code, we hope readers will be inspired to adopt HDR (if they haven't already done so) and then, through their endeavours, begin to move HDR on from the current limitations of "consumer HDR" to fulfill HDR's real potential for accurately imaging the real-world.

Advanced High Dynamic Range Imaging, 2nd edition covers all aspects of HDR imaging from capture to display, including an evaluation of just how closely the results of HDR processes are able to recreate the real world. In addition, this second edition provides an important focus on HDR video, which is of particular interest to the industry. The book is divided into 10 chapters. Chapter 1 introduces the basic concepts. This includes details on the way a human eye sees the world and how this may be represented digitally. Chapter 2 sets the scene for HDR imaging by describing the HDR pipeline and all that is necessary to capture real-world lighting and then subsequently display it. Chapters 3, 4, and 5 investigate the relationship between HDR and LDR content and displays. The numerous tone mapping techniques that have been proposed over more than 20 years are described in detail in Chapter 3. These methods tackle the problem of displaying HDR content in a desirable manner on LDR displays. Chapters 4 and Chapter 5 introduce advanced techniques for tone mapping images and videos. In Chapter 6, expansion operators, generally referred to as inverse (or reverse) tone mapping operators (iTMOs), are considered. This can be considered the opposite of tone mapping: how to expand LDR content for display on HDR devices. A major application of HDR technology, image-based lighting (IBL), is considered in Chapter 7. This computer graphics approach enables real and virtual objects to be relit by HDR lighting that has been previously captured. So, for example, the CAD model of a car may be lit by lighting previously captured in China to allow a car designer to consider how a particular paint scheme may appear in that country. Correctly applied IBL can thus allow such hypothesis testing without the need to take a physical car to China. Another example could be actors being lit accurately as if they were in places they have never been. Capturing real-world lighting generates a large amount of data. A single HDR frame of uncompressed 4K UHD resolution (3840 × 2160 pixels) requires approximately 95 Mbytes, and a minute of data at 30 fps is 167 Gbytes. Chapters 8 and 9 of *Advanced High Dynamic Range Imaging* examine the issues of compressing HDR imagery to enable it to be manageable for storage, transmission, and manipulation and thus practical on existing systems. Many tone mapping and expansion operators, and compression techniques have been proposed over the years. Several of these attempt to create as accurate a representation of the real world as possible within the constraints of the LDR display or content. Chapter 10 discusses methods that have been proposed to evaluate just how successful tone mappers have been in displaying HDR content on LDR devices and how successful expansion methods have been in recreating HDR images from legacy LDR content.

Introduction to MATLAB

MATLAB is a powerful numerical computing environment. Created in the late 1970s and subsequently commercialized by The MathWorks, MATLAB is now widely used across both academia and industry. The interactive nature of MATLAB allows it to rapidly demonstrate many algorithms in an intuitive manner. It is for this reason we have chosen to include the key HDR algorithms as MATLAB code as part of what we term the *HDR Toolbox*. An overview of the HDR Toolbox is given in Appendix C. In *Advanced High Dynamic Range Imaging*, the common parts of MATLAB code are presented at the beginning of each chapter. The remaining code for each technique is then presented at the point in the chapter where the technique is described. The code always starts with the input parameters that the specific method requires.

For example, in Listing 1, the code segment for the Schlick tone mapping operator [341] (see Section 3.2.3), the method takes the following parameters as input: s_mode specifies the type of model of the Schlick technique used.

```
%compute max luminance
LMax = MaxQuart(L, 0.99);
%comput min luminance
LMin = MaxQuart(L(L > 0.0), 0.01);

%mode selection
switch s_mode
    case 'manual'
        p = max([p, 1]);

    case 'automatic'
        p = L0 * LMax / (2^nBit * LMin);

    case 'nonuniform'
        p = L0 * LMax / (2^nBit * LMin);
        p = p * (1 - k + k * L / sqrt(LMax * LMin));
end

%dynamic range reduction
Ld = p .* L ./ ((p - 1) .* L + LMax);
```

Listing 1. MATLAB code: Schlick TMO [341].

There are three modes: manual, automatic, and nonuniform. The manual mode takes the parameter p as input from the user. The automatic and nonuniform modes use the uniform and non-uniform quantization technique, respectively. The variable p is either p or p' depending on the mode used, nBit is the number of bits N of the output display, L0 is L_0, and k is k. The first step is to compute the maximum, L_Max, and the minimum luminance, L_Min in a robust way (to avoid unpleasant appearances), i.e., computing percentiles. These values can be used for calculating p. Afterward, based on s_mode, one of the three modalities is chosen and either p is given by the user (manual mode) or is computed using Equation (3.9) (automatic mode) or using Equation (3.10) (nonuniform mode). Finally, the dynamic range of the luminance channel is reduced by applying Equation (3.8).

Acknowledgments

Many people provided help and support during the writing of this book. Special thanks go to the wonderful colleagues, staff, and professors I met during this time in Pisa, Warwick, Bristol, and around the world: Patrick, Kurt, Alessandro, Alan, Karol, Kadi, Luis Paulo, Sumanta, Piotr, Roger, Matt, Anna, Cathy, Yusef, Usama, Dave, Gav, Veronica, Timo, Alexa, Marina, Diego, Thomas Edward William, Jassim, Carlo, Elena, Alena, Belma, Selma, Jasminka, Vedad, Remi, Elmedin, Vibhor, Silvester, Gabriela, Nick, Mike, Giannis, Keith, Sandro, Georgina, Leigh, John, Paul, Mark, Joe, Gavin, Maximino, Alexandrino, Tim, Polly, Steve, Simon, Michael, Roberto, Matteo, Massimiliano C., Federico, Claudio, Frédéric L., Marco P., Claudia, Marco C., Marco T., Marco D.B., Giorgio, Giulia, Daniele B., Luca, Francisco, Alejandro, Victoria, Andrea, Francesca, Valeria, Eliana, Gaia, Gianpaolo, Nico, Luigi, Manuele, Oğuz, Ezgi, Ebru, Rosario, Guido, Rafał, Tania, Erik, Paul, Jonas, Anders, Per, Gabriel, Joel, Giuseppe V., Frédéric D., Francesco D.S., Rémi C., Olga, Daniele P., Tunç, Maurizio, Massimiliano, Sami, Gholi's sisters, Francesco P., Roberto M., Matteo C., Ricardo, Patrizio, and Orazio.

I am heavy with debt for the support I have received from my family. My parents, Maria Luisa and Renzo; my nieces, Agatha and Livia; my brother Piero and his wife, Irina; and my brother Paolo and his wife, Elisa.

—Francesco Banterle

I would like to thank my fiancé Despo, my fluffy white Lucia and all my family that have always supported my work. I would also like to thank with all my heart my mother, Franca, and grandmother Nella, who are always in my mind. Also a great thanks goes to all my co-authors on High Dynamic Range works and courses throughout the years.

—Alessandro Artusi

First, I am very grateful to the three co-authors whose hard work has made this book possible. I would like to thank the following colleagues, many of whom have been an inspiration and with whom it has been a pleasure working over the past few years at Bristol and Warwick: Vibhor Aggarwal, Matt Aranha, Ali Asadipour, Tom Bashford-Rogers, Max Bessa, Kadi Bouatouch, Keith Bugeja, Tim Bradley, Kirsten Cater, Joe Cordina, Gabriela Czanner, Silvester Czanner, Amar Dhokia, Efstratios Doukakis, Piotr Dubla, Sara de Freitas, Gavin Ellis, Anna Gaszczak, Jon Hatchett, Jassim Happa, Carlo Harvey, Vedad Hulusic, Emmanuel Ige, Richard Gillibrand, Brian Karr, Martin Kolář, Harjinder Lallie, Patrick Ledda, Pete Longhurst, Fotis Liarokapis, Cheng-Hung (Roger) Lo, Demetris Marnerides, Georgia Mastoropoulou, Josh McNamee, Miguel Melo, Christopher Moir, Ratnajit Mukharjee, Antonis Petroutsos, Alberto Proenca, Belma Ramic-Brkic, Selma Rizvic, Luis Paulo Santos, Pınar Satılmış, Simon Scarle, Elmedin Selmanovic, Debmalya Sinha, Sandro Spina, Rossella Suma, Kevin Vella, Greg Ward, and Xiaohui (Cathy) Yang. My parents have always supported me and I will be eternally grateful. My grandparents were an inspiration and are sorely missed—they will never be forgotten. Finally, I would like to wholeheartedly thank my wife, Anna, for her love and support and our children, Alex and Maya, who have made our lives complete.

—Kurt Debattista

This book provides an overview of the latest advances in HDR and has come about after more than 15 years of research in the field and a number of outstanding staff, post-docs and PhD students, three of whom are co-authors of this book. I am very grateful to all of them for their hard work over the years. This research has built on the work of the pioneers, such as Holly Rushmeier, Paul Debevec, Jack Tumblin, Helge Seetzen, Gerhard Bonnet and Greg Ward and together with the ever-increasing body of work from around the world has succeeded in taking HDR from a niche research area into general use. In the first edition of the book I predicted: "In the not too distant future, capturing and displaying real world lighting will be the norm with an HDR television in every home." While this is not quite true, yet, HDR has indeed reached the consumer market and HDR televisions can now be purchased. One of the key players in helping bring about the growth in interest in HDR was EU COST Action IC1005 which I had the honor to chair from 2011–2015. IC1005 helped coordinate HDR research and development across Europe and build a vibrant HDR community. In addition pioneering companies, such as Brightside and goHDR, paved the way for more established companies, including SIM2, Dolby, Technicolor, Philips, Sony, etc. to play leading roles in facilitating the widespread adoption of HDR. As HDR starts to enter all aspects of imaging, there will be many opportunities for additional innovative companies, such as our own TrueDR and Sensomatics, to emerge to introduce HDR to new fields, such as autonomous vehicles, security, imaging rocket launches, and virtual reality. In addition to all my Visualisation Group over the years, I would like to thank key collaborators and friends such as Christopher Moir, Trevor Power, Quentin Compton-Bishop, Pete Shirley, Maximino Bessa, Miguel Melo, Ali Asadapour, Brian Karr, Tom Bashford Rogers, Carlo Harvey, Luis Paulo dos Santos, Alexander Wilkie, Karol Myszkowski, Patrizio Campisi, George Papagiannakis, Rafał Mantiuk, Frédéric Dufaux, Giuseppe Valenzise, Pavel Zemčík, Andrej Ferko, Kadi Bouatouch, Domenico Toffoli and Igor G. Olaizola for their passion and enthusiasm for all things HDR over the years. Finally, thank you to Eva, Erika, Andrea, and Thomas for all their continued love and support.

—Alan Chalmers

MATLAB® is a registered trademark of The MathWorks, Inc. For product information please contact:

The MathWorks, Inc.
3 Apple Hill Drive
Natick, MA, 01760-2098 USA
Tel: (+1)508-647-7000
Fax: (+1)508-647-7001
E-mail: info@mathworks.com
Web: www.mathworks.com

1
Introduction

The computer graphics and related industries, in particular those involved with films, games, simulation, virtual reality, and military applications, continue to demand more realistic images displayed on a computer, that is, synthesized images that more accurately match the real scene they are intended to represent. This is particularly challenging when considering images of the natural world that present our visual system with a wide range of colors and intensities. A starlit night has an average luminance level of around 10^{-3} cd/m^2, and daylight scenes are close to 10^6 cd/m^2. Humans can see detail in regions that vary by 1:10^4 at any given eye adaptation level. With the possible exception of cinema, there has been little push for achieving greater dynamic range because common displays and viewing environments limit the range of what can be presented to about two orders of magnitude between minimum and maximum luminance. A well-designed cathode ray tube (CRT) monitor may do slightly better than this in a darkened room, but the maximum display luminance is only around 100 cd/m^2, and in the case of an LCD display the maximum luminance may reach 250–500 cd/m^2, which does not even begin to approach daylight levels. A high-quality xenon film projector may get a few times brighter than this, but it is still two orders of magnitude away from the optimal light level for human perception. This is now all changing with high dynamic range (HDR) imagery and novel capture and display HDR technologies, offering a step-change in traditional imaging approaches.

In the past two decades, HDR imaging has revolutionized the field of computer graphics and other areas such as photography, image processing, virtual reality, visual effects, and the video game industry. Real-world lighting can now be captured, stored, transmitted, and fully utilized for various applications without the need to linearize the signal and deal with clamped values. The very dark and bright areas of a scene can be recorded at the same time for both images and video, avoiding under-exposed and over-exposed areas; see Figure 1.1. Traditional imaging methods, on the other hand, do not use physical values and typically are constrained by limitations in technology that could only handle 8 bits per color channel per pixel. Such imagery (8 bits or less per color channel) is typically known as low dynamic range (LDR) imagery in computer graphics. In industry and standardization communities, i.e., JPEG and MPEG committees, it is commonly referred to as standard dynamic range (SDR) imagery.

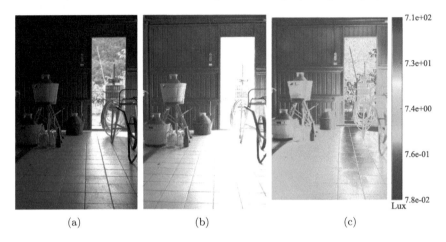

(a)   (b)   (c)

Figure 1.1. Images at different exposure times of the same scene that allow the capture of (a) very bright and (b) dark areas and (c) the corresponding HDR image in false colors.

An HDR image may be generated by capturing multiple images of the same scene at different exposure levels and merging them to reconstruct the original dynamic range of the captured scene. There are several algorithms for merging LDR images; Debevec and Malik's method [111] is an example of these. One commercial implementation is SpheronVR's SpheronCam HDR [351] that can capture still spherical images with a dynamic range of $6 \times 10^6 : 1$. Although information could be recorded in one shot using native HDR CCDs, problems of low sensor noise typically occur at high resolution.

HDR images/videos may occupy four times the amount of memory required by a corresponding LDR image. With HDR images, to cope with the full range of lighting, color values are stored using three floating point numbers. This has a major effect not only on storing and transmitting HDR data but also in terms of processing it. As a consequence, efficient representations of floating point numbers have been developed for HDR imaging, and many classic compression algorithms such as JPEG and MPEG have been adapted and extended to handle HDR images and videos.

Once HDR content has been efficiently captured and stored, it can be

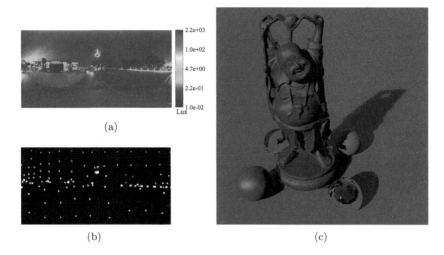

(a)

(b)   (c)

Figure 1.2. A relighting example: (a) An HDR environment image in false color. (b) Light sources extracted from it. (c) A relit Stanford's Happy Buddha model [154] using light sources generated in (b).

utilized for a variety of applications. One popular application is the relighting of synthetic or real objects. HDR data stores detailed lighting information of an environment. This information can be exploited for detecting light sources and using them for relighting objects; see Figure 1.2. Such relighting is very useful in many fields such as augmented reality, visual effects, and computer graphics. This is because the appearance of the image is transferred onto the relit objects producing synthetic images with high levels of realism.

Another important application of HDR is to capture samples of the bidirectional reflectance distribution function (BRDF), which describes how light interacts with a given material. These samples can be used to reconstruct the BRDF. HDR data is required for an accurate reconstruction; see Figure 1.3. In addition, HDR imaging may allow other applications, which use LDR images, to be more robust and accurate because more data is provided. For example, disparity calculations in computer vision can be improved in challenging scenes with bright light sources. This is because information in the light sources is not clamped; therefore, disparity can be computed for light sources and reflective objects with higher precision than when using clamped values [98].

Once HDR content is obtained, it needs to be visualized. HDR images and videos do not typically fit within the native dynamic range of classic LDR displays such as CRT or LCD monitors; e.g., a modern LCD monitor

(a)   (b)

Figure 1.3. An example of capturing different BRDFs of a Prun stone [132]: (a) A setup for capturing BRDFs using HDR technology inspired by Tchou et al. [365]. (b) The reconstructed materials in (a) from 64 samples for each of three exposures.

has a dynamic range between 1 and 250 cd/m^2. Therefore, when using such displays, the HDR content has to be processed to match the available dynamic range of the display. This operation is called tone mapping and it is introduced in Chapter 3; see Figure 1.4. Seetzen et al. [342] proposed the first solutions for HDR displays that can natively visualize HDR content; Chapter 2 gives an overview on this topic.

(a)   (b)

Figure 1.4. An example of HDR visualization on an LDR monitor of the Cellar HDR image: (a) The HDR image in false color. (b) The image in (a) has been processed to visualize details in bright and dark areas. This process is called tone mapping.

1.1 Light, Human Vision, and Color Spaces

This section introduces basic concepts of visible light, the units used for measuring it, the human visual system (HVS) focusing on the eye, and color spaces. These concepts are fundamental in HDR imaging as they encapsulate the physical-real values of light, from very dark values (i.e., 10^{-3} cd/m^2) to very bright ones (i.e., 10^6 cd/m^2). The perception of a scene depends greatly on the lighting conditions.

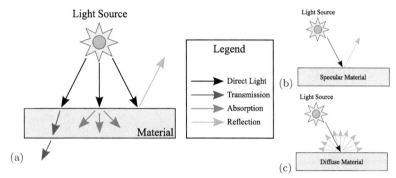

Figure 1.5. (a) The three main light interactions: transmission, absorption, and reflection. In transmission, light travels through the material, changing its direction according to the physical properties of the medium. In absorption, the light is taken up by the material that was hit and is converted into thermal energy. In reflections, light bounces from the material in a different direction due to the material's properties. There are two main kinds of reflections: specular and diffuse. (b) Specular reflections: an incoming ray is reflected in a particular outgoing direction depending on the material's optical properties. (c) Diffuse reflections: an incoming ray is reflected in all directions with equal energy (ideally).

1.1.1 Light

Visible light is a form of radiant energy that travels in space, interacting with materials where it can be absorbed, refracted, reflected, and transmitted; see Figure 1.5. Traveling light can reach human eyes, stimulating them to produce visual sensations depending on the wavelength; see Figure 1.6.

Radiometry and photometry define how to measure light and its units over time, space, and direction. While the former measures physical units, the latter takes into account the human eye, where spectral values are weighted by the spectral response $\bar{y}$ of a standard observer; see Figure 1.8. Radiometric and photometric units were standardized by the Commission Internationale de l'Eclairage (CIE) [93]. The main radiometric units are

- Radiant energy (Ω_e). This is the basic unit for light. It is measured in joules (J).

- Radiant power ($P_e = \frac{\Omega_e}{dt}$). Radiant power is the amount of energy that flows per unit of time. It is measured in watts (W); $W = J \times s^{-1}$.

- Radiant intensity ($I_e = \frac{dP_e}{d\omega}$). This is the amount of radiant power per unit of direction. It is measured in watts per steradian (W × sr^{-1}).

- Irradiance ($E_e = \frac{dP_e}{dA_e}$). Irradiance is the amount of radiant power per unit of area from all directions of the hemisphere at a point. It is measured in watts per square meters (W × m^{-2}).

- Radiance ($L_e = \frac{d^2 P_e}{dA_e \cos\theta d\omega}$). Radiance is the amount of radiant power arriving/leaving at a point in a particular direction. It is measured in watts per steradian per square meter (W × sr^{-1} × m^{-2}).

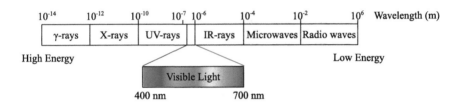

Figure 1.6. The electromagnetic spectrum. The visible light covers a very limited spectrum between 400 nm and 700 nm of the full electromagnetic spectrum.

The main photometric units are

- Luminous power (P_v). Luminous power is the weighted radiant power. It is measured in lumens (lm), a derived unit from candela (lm = cd × sr).

- Luminous energy (Q_v). This is analogous to the radiant energy. It is measured in lumens per second (lm × s^{-1}).

- Luminous intensity (I_v). This is the luminous power per direction. It is measured in candela (cd), which is equivalent to lm × sr^{-1}.

- Illuminance (E_v). Illuminance is analogous to irradiance. It is measured in lux, which is equivalent to lm × m^{-2}.

- Luminance (L_v). Luminance is the weighted radiance. It is measured in cd × m^{-2} or nit; it is equivalent to lm × m^{-2} × sr^{-1}.

A measure of the relative luminance of the scene can be useful, since it can illustrate some properties of the scene such as the presence of diffuse or specular surfaces, lighting condition, etc. For example, specular surfaces reflect light sources even if they are not visible directly in the scene, increasing the relative luminance. This relative measure is called *contrast*. Contrast is formally defined as the relationship between the darkest and the brightest value in a scene, and it can be calculated in different ways. The main contrast relationships are Weber Contrast (C_W), Michelson Contrast (C_M), and Ratio Contrast (C_R). These are defined as

$$C_W = \frac{L_{\max} - L_{\min}}{L_{\min}}, \qquad C_M = \frac{L_{\max} - L_{\min}}{L_{\max} + L_{\min}}, \qquad C_R = \frac{L_{\max}}{L_{\min}},$$

where $L_{\min}$ and $L_{\max}$ are, respectively, the minimum and maximum luminance values of the scene. Throughout this book C_R is used as the generalized definition of contrast.

1.1.2 An Introduction to the Human Eye

The eye is an organ that gathers light onto photoreceptors, which then convert light into signals; see Figure 1.7. These are transmitted through the optical nerve to the visual cortex, an area of the brain that processes these signals producing the perceived image. This full system, which is responsible for vision, is referred to as the HVS [254].

Light, which enters the eye, first passes through the cornea, a transparent membrane. Then, it enters the pupil, an aperture that is modified by the iris, a muscular diaphragm. Subsequently, light is refracted by the lens and hits photoreceptors in the retina. Inside the eye there are two liquids, the vitreous and aqueous humors. The former fills the eye, keeping its shape and the retina against the inner wall. The latter is between the cornea and the lens and maintains the intraocular pressure [254].

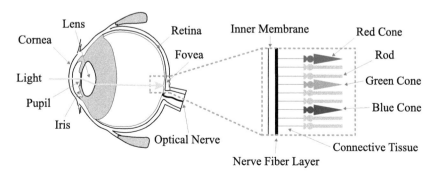

Figure 1.7. The human eye.

There are two types of photoreceptors: cones and rods. There are approximately 6–7 million cones located mostly in the fovea [131]. The spectral sensitivities of the photoreceptor rod and cone systems at different luminance levels are described by the *scotopic, mesopic,* and *photopic* luminous efficiency functions. These are also called contrast sensitivity functions (CSFs). The cones are sensitive at luminance levels between 10^{-2} cd/m^2 and 10^8 cd/m^2 (photopic vision or daylight vision), and are responsible for the perception of high frequency patterns, fast motion, and colors. Furthermore, color vision is due to three types of cones: short wavelength cones, sensitive to wavelengths around 435 nm; middle wavelength cones, sensitive around 530 nm; and long wavelength cones, sensitive around 580 nm. The rods, of which there are 75–150 million [131], are sensitive at luminance levels between 10^{-6} cd/m^2 and 10 cd/m^2 (scotopic vision or night vision). In this range, the rods are more sensitive than cones but do not provide color vision. This is the reason why we are unable to discriminate between colors at low level illumination conditions. There is only one type of rod and it is located around the fovea but is absent in it. This is why high frequency patterns cannot be distinguished at low lighting conditions. The mesopic range, where both rods and cones are active, is defined between 10^{-2} cd/m^2 and 10 cd/m^2. Note that an adaptation time is needed for passing from photopic to scotopic vision and vice versa; for more details; see [254]. The rods and cones compress the original signal, reducing the dynamic range of incoming light. This compression follows a sigmoid function:

$$\frac{R}{R_{\max}} = \frac{I^n}{I^n + \sigma^n}, \tag{1.1}$$

where R is the photoreceptor response, $R_{\max}$ is the maximum photoreceptor response, and I is the light intensity. The variables σ and n are, respectively, the semisaturation constant and the sensitivity control exponent, which are different for cones and rods [254]. Another important aspects is the so-called visual threshold, which is the probability of seeing a visual difference in any visual attributes exhibited by a given set of stimuli [422]. Curves known as threshold vs. intensity (TVI) functions have been derived by experimental data, and they determine the relationship between the threshold and the background luminance's, L. This relationship is described by *Weber's law*, which provides a linear behavior on a log-log scale, and it is defined as

$$\Delta L = k \cdot L, \tag{1.2}$$

where ΔL is the smallest detectable luminance difference, and k is constant. Equation (1.2) states that the HVS has constant contrast sensitivity since

Weber's contrast, $\Delta L/L = k$, is constant over this range [301]. Furthermore, Equation (1.2) highlights that adding or subtracting luminance to a stimulus will have different visual impact depending on the background luminance.

1.1.3 Color Spaces

A color space is a mathematical description for representing colors, typically represented by three components called primary colors. There are two classes of color spaces: device dependent and device independent. The former describes the color information in relation to the technology used by the color device to reproduce the color. In the case of a computer monitor it depends on the set of primary phosphors, while in an ink-jet printer it depends on the set of primary inks. A drawback of this representation is that a color with the same coordinates such as $R = 150$, $G = 40$, $B = 180$ will not appear the same when represented on different monitors. The device independent class is not dependent on the characteristics of a particular device. In this way a color represented in such a color space always corresponds to the same color information. A typical device-dependent color space is the RGB color space. The RGB color space is a Cartesian cube represented by three additive primaries: red (R), green (G), and blue (B). A typical independent color space is the CIE 1931 XYZ color space, which is formally defined as the projection of a spectral power distribution I into the color-matching functions, $\overline{x}$, $\overline{y}$, and $\overline{z}$:

$$X = \int_{380}^{830} I(\lambda)\overline{x}(\lambda)d\lambda, \quad Y = \int_{380}^{830} I(\lambda)\overline{y}(\lambda)d\lambda, \quad Z = \int_{380}^{830} I(\lambda)\overline{z}(\lambda)d\lambda,$$

where λ is the wavelength.

Figure 1.8 shows plots of $\overline{x}$, $\overline{y}$, and $\overline{z}$. Note that the XYZ color space was designed in such a way that the Y component measures the luminance of the color. The chromaticity of the color is derived from XYZ values as

$$x = \frac{X}{X + Y + Z}, \quad y = \frac{Y}{X + Y + Z}.$$

These values can be plotted, producing the so-called CIE xy chromaticity diagram. This diagram shows all colors perceivable by the HVS; see Figure 1.8(b).

A popular color space for CRT and LCD monitors is the $sRGB$ color space [355], which uses the primaries of ITU-R Recommendation BT.709 [383] (or BT.709) for high-definition (HD) television. This color space defines as primaries the colors red (R), green (G), and blue (B). Furthermore, each color in $sRGB$ is a linear additive combination of values in $[0, 1]$ of the

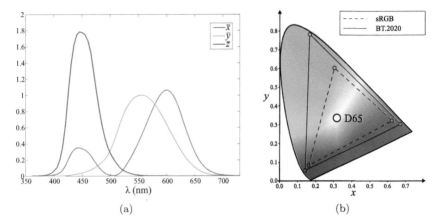

(a)   (b)

Figure 1.8. The CIE XYZ color space: (a) The CIE 1931 two-degree XYZ color matching functions. (b) The CIE xy chromaticity diagram showing all colors that HVS can perceive. The small triangle (dashed line) is the space of colors that can be represented in $sRGB$, where the three circles represent the three primaries. The large triangle (continuous line) is the space of colors that can be represented in BT.2020. This color space has a larger gamut than $sRGB$.

three primaries. Therefore, not all colors can be represented, only those inside the triangle generated by the three primaries; see Figure 1.8(b).

A linear relationship exists between the XYZ and RGB color spaces:

$$\begin{bmatrix} X \\ Y \\ Z \end{bmatrix} = \mathbf{M} \begin{bmatrix} R \\ G \\ B \end{bmatrix}.$$

In the case of $sRGB$, $\mathbf{M}$ is defined as

$$\mathbf{M}_{sRGB} = \begin{bmatrix} 0.412 & 0.358 & 0.181 \\ 0.213 & 0.715 & 0.072 \\ 0.019 & 0.119 & 0.950 \end{bmatrix}.$$

$sRGB$ presents a non-linear transformation for each color channel to linearize the signal when displayed on LCD and CRT display. Such a nonlinear transformation is called an electro-optical transfer function (EOTF) in the standards community. Its inverse is typically called an opto-electrical transfer function (OETF). This is because there is a non-linear relationship between the output intensity generated by the displaying device and the input voltage. This relationship is modeled as

$$C_v = \begin{cases} 12.92C & \text{if } C \leq 0.0031308, \\ 1.055C^{\frac{1}{\gamma}} - 0.055 & \text{otherwise,} \end{cases}$$

where C is a color channel, $\gamma = 2.4$, and the subscript $_v$ means that the color channel is ready for visualization. Note that this equation has been derived from the gamma correction:

$$C_v = C^{\frac{1}{\gamma}},$$

where γ is typically in the range $[1.8, 2.4]$.

An emerging color space for ultra-high definition television (UHDTV) is ITU-R Recommendation BT.2020 [382] (or BT.2020). This color space has a wider gamut than $sRGB$ but it cannot represent the full gamut of colors; see Figure 1.8(b). In the case of BT.2020, $\mathbf{M}$ is defined as

$$\mathbf{M}_{\text{BT.2020}} = \begin{bmatrix} 0.636958 & 0.144617 & 0.168881 \\ 0.262700 & 0.677998 & 0.059302 \\ 0.000000 & 0.028073 & 1.060985 \end{bmatrix}.$$

Furthermore, a similar EOTF is defined as

$$C_v = \begin{cases} 4.5C & \text{if } C \leq \beta, \\ \alpha C^{\frac{1}{\gamma}} - (\alpha - 1) & \text{otherwise,} \end{cases}$$

where $\gamma = 2.2$, $\alpha = 1.099$ and $\beta = 0.018$ for 10-bit systems, and $\alpha = 1.0993$ and $\beta = 0.0181$ for 12-bit systems.

Symbol	Description
L_{w}	HDR luminance value
L_{d}	LDR luminance value
L_{avg}	Arithmetic mean luminance value
L_{H}	Logarithmic mean luminance value
L_{min}	Minimum luminance value
L_{max}	Maximum luminance value

Table 1.1. The main symbols used for the luminance channel in HDR image processing.

The RGB color space is popular in HDR imaging; other practical color spaces are presented in Appendix B. However, many computations are calculated in the luminance channel Y from XYZ, which is usually referred to as L. In addition, common statistics from this luminance are often used, such as the maximum value, L_{max}, the minimum one, L_{min}, and the mean value. This can be computed as the arithmetic average, L_{avg} or the logarithmic one, L_{H}:

$$L_{\text{avg}} = \frac{1}{N} \sum_{i=1}^{N} L(\mathbf{x}_i), \qquad L_{\text{H}} = \exp\left(\frac{1}{N} \sum_{i=1}^{N} \log\big(L(\mathbf{x}_i) + \epsilon\big) \right),$$

where $\mathbf{x}_i = (x_i, y_i)^\top$ are the coordinates of the i-th pixel, and $\epsilon > 0$ is a small constant for avoiding singularities, i.e., $\log 0$. Note that in HDR imaging, subscripts $_\mathrm{w}$ and $_\mathrm{d}$ (representing world luminance and display luminance, respectively) refer to HDR and LDR values. Table 1.1 shows the main symbols used in HDR image processing for the luminance channel.

2
HDR Pipeline

HDR imaging is a revolution in the field of imaging permitting the ability to use and manipulate physically real light values. This chapter introduces the main processes of HDR imaging, which can be best characterized as a pipeline, termed the HDR pipeline. Figure 2.1 illustrates the distinct stages of the HDR pipeline.

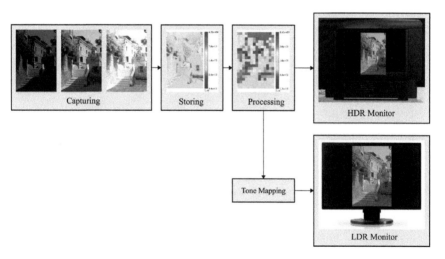

Figure 2.1. The HDR pipeline in all its stages. Multiple exposure images are captured and combined, obtaining an HDR image. Then this image is quantized, compressed, and stored. Further processing can be applied to the image. For example, areas of high luminance can be extracted and used to relight a synthetic object. Finally, the HDR image or a tone mapped HDR image can be visualized using native HDR monitors or traditional LDR displays.

The first stage concerns the generation of HDR content. HDR content can be captured in a number of ways, although limitations in hardware technology, until recently, have meant that HDR content capture has typically required the assistance of software. Section 2.1 outlines different ways in which HDR images can be generated. These include images generated from a series of still LDR images, using computer graphics, and via expansion from single-exposure images. The section also describes exciting new hardware that enables native HDR capture.

Due to the explicit nature of HDR values, HDR content may be considerably larger than its LDR counterpart. To make HDR manageable, efficient storage methods are necessary. In Section 2.2, HDR file formats are introduced. Compression methods can also be applied at this stage; Chapter 8 discusses them in detail.

Finally, HDR content can be natively visualized using a number of new display technologies. Section 2.3 introduces the primary native HDR displays. Such displays have become available to consumers. However, software solutions can be employed to adapt HDR content to be shown on LDR displays while attempting to maintain an HDR viewing experience. Such software solutions take the form of operators that transform the range of luminance in the HDR image to the luminance range of the LDR display. These operators are termed tone mapping operators (TMOs) and a large variety of tone mapping operators exist. Chapter 3 introduces tone mapping in detail.

2.1 Acquisition of HDR Content

In this book, four methods of generating HDR content are presented. The first, and most commonly used until recently, is the generation of HDR content by combining a number of LDR captures at different exposures through the use of software technology. The second, which has become more feasible, is the direct capture of HDR images using specialized hardware. The third method, popular in the entertainment industries, is the creation of HDR content from virtual environments using physically based renderers. The final method is the generation of HDR content from legacy content consisting of single exposure captures, using software to expand the dynamic range of the LDR content or reference HDR images.

2.1.1 Capturing HDR Images by Combining Multiple Exposures

At the time of writing, available consumer cameras and videocameras are limited since they can only capture 8-bit images or 14-bit images in RAW format. This does not cover the full dynamic range of irradiance values in most environments in the real world.

(a)   (b)   (c)   (d)

Figure 2.2. An example of HDR capturing for generating the Dorsoduro HDR image. Images taken with different exposure times: (a) $\frac{1}{500}$ sec. (b) $\frac{1}{125}$ sec. (c) $\frac{1}{30}$ sec. The HDR image is obtained by combining (a), (b), and (c). (d) A false color rendering of the luminance channel of the obtained HDR image.

The most commonly used method of capturing HDR images is to acquire multiple photographs of the same scene at different exposure times to capture all details from the darkest to the brightest areas; see Figure 2.2. This set of images is typically called an *exposure stack*. If the camera has a linear response, the irradiance, E, at each pixel location, $\mathbf{x}$, can be recovered by combining all recorded values at different exposure times as

$$E(\mathbf{x}) = \frac{\sum_{i=1}^{N_e} w(I_i(\mathbf{x})) \frac{I_i(\mathbf{x})}{\Delta t_i}}{\sum_{i=1}^{N_e} w(I_i(\mathbf{x}))}, \tag{2.1}$$

where I_i is the image recording linear sensor's values at the i-th exposure, Δt_i is the exposure time for I_i, and N_e is the number of images at different exposures. $w(I_i(\mathbf{x}))$ is a weighting function that removes outliers: high values in one of the exposures will have less noise than low values; on the other hand, high values can be saturated. Typically, middle values are more reliable. Figure 2.3 shows a few examples of w functions. Figure 2.2(d) shows an example of the recovered irradiance map using Equation (2.1). Note that Equation (2.1) computes sensor irradiance and not scene radiance. Typically, L is proportional to E. Although modern lenses try to maintain a constant mapping by design, some areas of the sensor may exhibit a different mapping. One way to correct this is given by [172]:

$$L = \frac{E}{\frac{\pi}{4} \cos^4 \alpha} \left(\frac{R}{d} \right)^2 \tag{2.2}$$

where α measures the pixel's angle from the lens' optical axis, d is the distance between the lens and the image plane, and R is the radius of the lens.

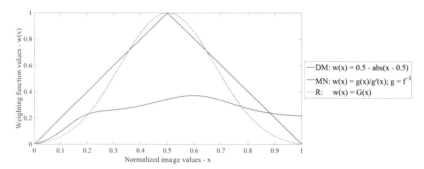

Figure 2.3. Different w functions for weighting different exposures: DM (blue) and R (yellow) were, respectively, proposed by Debevec and Malik [111] and Robertson et al. [331] to use central values, which are thought to be more reliable. MN (red) was proposed by Mitsunaga and Nayar [271] to maximize both SNR and sensitivity to radiance changes of a sensor. Note that G is a Gaussian function with $\mu = 0.5$ and $\sigma = 1/\sqrt{32}$ that is shifted and normalized.

Although some manufacturers may offer the possibility to capture content in RAW formats, which store values linearly, film and digital cameras do not typically have a linear response but a more general function f, called the camera response function (CRF); see Figure 2.4.

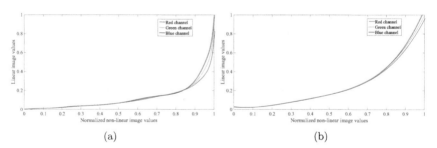

Figure 2.4. An example of estimated inverse CRF of a Canon 550D using two different methods: (a) Debevec and Malik's method [111]. (b) Mitsunaga and Nayar's method [271].

The CRF attempts to compress as much of the dynamic range of the real world as possible into the limited 8/10-bit storage or into film medium. Therefore, Equation (2.1) needs to be updated to take this characteristic into account:

$$E(\mathbf{x}) = \frac{\sum_{i=1}^{N_e} w(I_i(\mathbf{x})) \frac{f^{-1}(I_i(\mathbf{x}))}{\Delta t_i}}{\sum_{i=1}^{N_e} w(I_i(\mathbf{x}))}, \qquad (2.3)$$

where f^{-1} is the inverse CRF, which is what we are interested in estimating;

see Figure 2.4. Note that Equation (2.3), due to numerical precision errors, can produce very bright isolated pixels, i.e., salt and pepper noise. To reduce these pixels, Equation (2.3) needs to be computed either in the log-domain [111] or the influence of very dark exposures that contain very noisy pixels needs to be reduced [332]. In the first case, Equation (2.3) becomes:

$$E(\mathbf{x}) = \exp\left(\frac{\sum_{i=1}^{N_e} w(I_i(\mathbf{x})) \log f^{-1}(I_i(\mathbf{x})) - \log \Delta t_i}{\sum_{i=1}^{N_e} w(I_i(\mathbf{x}))}\right). \qquad (2.4)$$

A straightforward way to reduce the influence of dark pixels in Equation (2.3) is to scale w by Δt_i^2, i.e., the w has less influence at low exposure time, obtaining:

$$E(\mathbf{x}) = \frac{\sum_{i=1}^{N_e} w(I_i(\mathbf{x})) f^{-1}(I_i(\mathbf{x})) \Delta t_i}{\sum_{i=1}^{N_e} w(I_i(\mathbf{x})) \Delta t_i^2}, \qquad (2.5)$$

which is called squared domain merge. Note that more sophisticated methods can be employed based on filtering [16], or capturing the optimal sub-set of images [164] maximizing SNR.

```
%this value is added for numerical stability
delta_value = 1.0 / 65536.0;

%for each LDR image...
for i=1:n
    tmpStack = ClampImg(single(stack(:,:,:,i)) / scale, 0.0,
        1.0);

    %computing the weight function
    weight   = WeightFunction(tmpStack, weight_type, bMeanWeight
        );

    %linearization of the image
    tmpStack = RemoveCRF(tmpStack, lin_type, lin_fun);

    %sum things up...
    dt_i = stack_exposure(i);

    switch merge_type
        case 'linear'
            imgOut = imgOut + (weight .* tmpStack) / dt_i;
            totWeight = totWeight + weight;

        case 'log'
            imgOut = imgOut + weight .* (log(tmpStack +
                delta_value) - log(dt_i));
            totWeight = totWeight + weight;

        case 'w_time_sq'
```

```
            imgOut = imgOut + (weight .* tmpStack) * dt_i;
            totWeight = totWeight + weight * dt_i * dt_i;
      end
end

imgOut = (imgOut ./ totWeight);

if(strcmp(merge_type, 'log') == 1)
    imgOut = exp(imgOut);
end
```

Listing 2.1. MATLAB code: Combining multiple LDR images into an HDR image.

Listing 2.1 shows MATLAB® code for combining multiple LDR images at different exposure times into a single HDR image. The full code is given in the file BuildHDR.m. After handling the input parameters, the main function sums up the contribution of each linearized image, using the function RemoveCRF.m, in the stack. Each contribution is weighted using the function WeightFunction.m, and it can be added in either the linear domain, or logarithm domain, or squared domain depending on merge_type (an input parameter).

Estimating a CRF. Mann and Picard [239] proposed a simple method for calculating f, which consists of fitting the values of pixels at different exposure to a fixed $f(x) = ax^\gamma + b$. This parametric function f is limited and does not support most real CRFs.

Debevec and Malik [111] proposed a straightforward and general method for recovering a CRF. For the sake of clarity, this method and others will be presented for gray channel images; in the case of color images, the method needs to be applied to each color channel. The value of a pixel in an image is given by the application of a CRF to the irradiance scaled by the exposure time:

$$I(\mathbf{x}) = f(E(\mathbf{x})\Delta t_i).$$

Rearranging terms and applying a logarithm to both sides, this becomes:

$$\log(f^{-1}(I(\mathbf{x}))) = \log E_i(\mathbf{x}) + \log \Delta t_i. \qquad (2.6)$$

Assuming that f is a smooth and monotonically increasing function, f and E can be calculated by minimizing the least square error derived from

Equation (2.6) using pixels from images at different exposures as

$$\mathcal{O} = \sum_{i=1}^{N_e} \sum_{j=1}^{M} \left(w\big(I_i(\mathbf{x}_j)\big) \big[g(I_i(\mathbf{x}_j)) - \log E(\mathbf{x}_j) - \log \Delta t_i\big] \right)^2$$

$$+ \lambda \sum_{x=T_{\min}+1}^{T_{\max}-1} (w(x)g''(x))^2, \tag{2.7}$$

where $g = f^{-1}$ is the inverse of the CRF, M is the number of pixels used in the minimization, and $T_{\max}$ and $T_{\min}$ are, respectively, the maximum and minimum integer values in all images I_i. The second part of Equation (2.7) is a smoothing term for removing noise, where function w is defined as

$$w(x) = \begin{cases} x - T_{\min} & \text{if } x \leq \frac{1}{2}(T_{\max} + T_{\min}), \\ T_{\max} - x & \text{if } x > \frac{1}{2}(T_{\max} + T_{\min}). \end{cases}$$

Mitsunaga and Nayar [271] proposed a polynomial representation of f^{-1} that can be defined as

$$f^{-1}(x) = \sum_{k=0}^{P} c_k x^k.$$

At this point the calibration process can be reduced to the estimation of the polynomial order P and the coefficients c_j. Taking two images of a scene with two different exposure times Δt_1 and Δt_2, the ratio R can be written as

$$R = \frac{\Delta t_1}{\Delta t_2} = \frac{I_1(\mathbf{x})}{I_2(\mathbf{x})}. \tag{2.8}$$

The brightness measurement $I_i(\mathbf{x})$ produced by an imaging system is related to scene radiance $L(\mathbf{x}) \propto E(\mathbf{x})$ for the time interval Δt_i via a response function $I_i(\mathbf{x}) = f(E(\mathbf{x}\Delta t_i))$. From this, $I_i(\mathbf{x})$ can be rewritten as $E(\mathbf{x}\Delta t_i) = g(I_i(\mathbf{x}))$ where $g = f^{-1}$. Since the response function of an imaging system is related to the exposure ratio, Equation (2.8) can be rewritten as

$$R_{1,2}(\mathbf{x}) = \frac{I_1(\mathbf{x})}{I_2(\mathbf{x})} = \frac{\sum_{k=0}^{P} c_k I_1(\mathbf{x})^k}{\sum_{k=0}^{P} c_k I_2(\mathbf{x})^k}, \tag{2.9}$$

where the images are ordered in such a way that $\Delta t_1 < \Delta t_2$ so that $R \in (0,1)$. The number of $f - R$ pairs that satisfy Equation (2.9) is infinite. This ambiguity is alleviated by the use of the polynomial model. The response function can be recovered by formulating the error function as

$$\varepsilon = \sum_{i=1}^{N_e} \sum_{j=1}^{M} \left(\sum_{k=0}^{P} c_k I_i(\mathbf{x}_j)^k - R_{i,i+1}(\mathbf{x}_j) \sum_{k=0}^{P} c_k I_{i+1}(\mathbf{x}_j)^k \right)^2, \tag{2.10}$$

where all measurements can be normalized so that $I_i(\mathbf{x})$ is in $[0, 1]$. An additional constraint can be introduced if the indeterminable scale can be fixed as $f(1) = I_{\max}$, which follows $c_P = I_{\max} - \sum_{k=0}^{P-1} c_k$. The coefficients of the CRF are determined by solving a linear system setting:

$$\frac{\partial \varepsilon}{\partial c_k} = 0.$$

After analyzing 201 CRFs from real-world cameras, Grossberg and Nayar [153] noticed that real-world CRFs occupy a small part of the theoretical space of all possible CRFs. Therefore, they proposed an inverse CRF model base on a basis function:

$$f^{-1}(x) = h_0(x) + \sum_{k=1}^{P} c_k h_k(x),$$

where $\{h_1, ..., h_P\}$ are the basis functions, which are computed by principal components analysis (PCA) on the 201 real-world CRFs, $\{c_1, ..., c_P\}$ are the coefficients of the model, and h_0 is the mean inverse CRF. The c_k coefficients of the model are computed using an error function similar to Equation (2.10).

A different approach was proposed by Robertson et al. [331,332]. Their method estimates the unknown inverse response function as well as the irradiance, $E(\mathbf{x})$, through the use of the maximum likelihood approach, where the objective function to be minimized is

$$\mathcal{O}(I, E) = \sum_{i=0}^{N_e} \sum_{j=0}^{M} w \left(I_i(\mathbf{x}_j) - \Delta t_i E(\mathbf{x}_j)\right)^2,$$

where w is a Gaussian function, which represents the noise in the imaging system used to capture the images.

Which inverse CRF estimation method to use? Among all different methods for recovering the inverse CRF from a stack of LDR images at different exposures, it may be difficult to identify which is best to use. Akyüz and Gençtav [294] investigated this matter, and their overall suggestions are: Debevec and Malik's algorithm [271] and Mitusnaga and Nayar are the most robust to noise, Grossberg and Nayar's method [153] is the most accurate, and Mitsunaga and Nayar [271] is the most consistent. With regards to accuracy, there is no particular winner. Another important result of this work is that either too dark or too bright scenes are not suitable when recovering inverse CRF.

Handling colors. Note that all the presented methods for recovering the inverse CRF can be extended to color images applying each method separately for each color band. This is because each color inverse CRF could be in theory independent from the others. However, when computing three different normalized inverse CRFs the relative scalings are unknown. A straightforward but effective method to determine these scalings, $\mathbf{k} = [k_R, k_G, k_B]$ (in the case of RGB), is to preserve the chromaticity of scene points [271] as a minimization process:

$$\arg\min_{\mathbf{k}} \sum_{c \in \{R,G,B\}} \left(\frac{\mathbf{Z}_c}{\|\mathbf{Z}\|} - \frac{(\mathbf{X} \odot \mathbf{k})_c}{\|\mathbf{X} \odot \mathbf{k}\|} \right)^2 \qquad X_c = f_c^{-1}(Z_c), \qquad (2.11)$$

where $\odot$ is the element wise product, and $\mathbf{Z}$ is a color value, which is typically set to $[0.5, 0.5, 0.5]$ to map the mid-gray to equal radiance values for R, G, and B [111].

Handling outliers during merge. An exposure stack can have over-exposed pixels, i.e., values close to 1 (assuming normalized values in $[0, 1]$), and under-exposed pixels, i.e., values close to 0, when not all needed exposures are acquired. In these cases, the normalization term

$$N(\mathbf{x}) = \sum_{i=1}^{N_e} w(I_i(\mathbf{x})),$$

in all different variations of Equation (2.1) can lead to a singularity. A solution is to clamp this value to a small value $\epsilon > 0$, i.e., $N = \max(N, \epsilon)$. However, this may be biased. In the case of over-exposed pixels, a possible solution is to determine the k-th exposure with the shortest exposure time in the stack and to compute E as

$$E(\mathbf{x}) = \frac{f^{-1}(I_k(\mathbf{x}))}{\Delta t_k} \quad \text{if } N(\mathbf{x}) < \epsilon.$$

This process can also be applied similarly to under-exposed pixels.

```
%checking for singularities
threshold = 1e-4;
if(~isempty(totWeight <= threshold))
    i_med = round(length(stack_exposure) / 2);
    med = ClampImg(single(stack(:,:,:,i_med)) / scale, 0.0,
        1.0);
    %over-exposed pixels
    [~, i_sat] = max(stack_exposure);
    tmpStack = ClampImg(single(stack(:,:,:,i_sat)) / scale,
        0.0, 1.0);
    img_sat = RemoveCRF(tmpStack, lin_type, lin_fun);
```

```
    img_sat = img_sat / stack_exposure(i_sat);

    mask = zeros(size(totWeight));
    mask(totWeight <= saturation & med > 0.5) = 1;
    mask = max(mask, [], 3);

    for i=1:col
        tmp = imgOut(:,:,i);
        slice_i = img_sat(:,:,i);
        tmp(mask == 1) = slice_i(mask == 1);
        imgOut(:,:,i) = tmp;
    end
    %under-exposed pixels
    ...
end
```

Listing 2.2. MATLAB code: Handling saturated values in the assembled HDR image.

Listing 2.2 shows MATLAB code for removing singularities when merging LDR images at different exposure times. The full code is given in the file BuildHDR.m. If totWeight is less than a given threshold (a small positive value) there may be a singularity. In this case, all affected pixels in mask are removed and replaced with linearized pixels from the shortest exposure, img_sat divided by its exposure time, stack_exposure(i_sat).

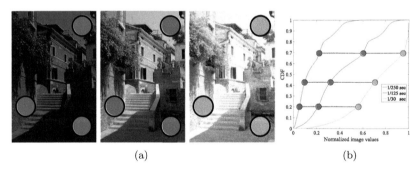

(a)   (b)

Figure 2.5. Different strategies for drawing samples for the CRF estimation: (a) Spatial sampling; after a spatial location, **x** is computed (e.g., regular sampling, random sampling, etc.), a sample is drawn by collecting LDR image values at different exposure times at **x**. (b) Histogram sampling; a sample is drawn by collecting the normalized image values of intersections between the horizontal line (dotted) at a given percentile (which can be regular, random, etc.) and CDFs of the LDR images histograms.

How to choose M. If all pixels were employed for computing the inverse CRF this would lead to relatively slow computation. Moreover, the use of a huge amount of data could create numerical instabilities when solving the optimization processes. One of the first solutions introduced was to apply subsampling either regular or irregular; see Figure 2.5(a). However, this methodology is sub-optimal because important samples or sampling areas may be missed, and all images in the stack need to be perfectly aligned unless feature matching is employed [226]. Grossberg and Nayar [152] proposed to draw samples from the CDF histograms of images in the stack. The corresponding camera sensor's values at different exposure times are found at the same percentile value; see Figure 2.5(b). This solution is robust to moderate camera movements and people/objects movement in the scene. For large motions, a RANSAC-based strategy can be employed [35] using rank-minimization for radiometric calibration [221] to avoid overfitting.

Calibration. Cameras are not measuring devices, they are meant for taking a shot of life blessed by light. Therefore, camera values are not absolute in terms of cd/m^2 or matching real-world colors. A simple approach for estimating absolute luminance values is to compute the average scene luminance of the i-th shot, $L_{i,\text{avg}}$ as

$$L_{i,\text{avg}} = K \frac{A_i^2}{\Delta t_i S_i},\qquad(2.12)$$

where A_i is its relative aperture (f-number), Δt is its exposure time (secs), S_i is its ISO arithmetic speed, and K is a calibration constant depending on the camera, which varies between 10.6 and 13.4 [133]. From Equation (2.12), the exposure value, e_i, of the i-th image to be used instead of Δt_i can be computed as

$$e_i = \frac{S_i \Delta t_i}{K A_i^2}.\qquad(2.13)$$

A more accurate process is to measure a few areas of the scene to capture with a luminance meter. However, this device is quite expensive, and satisfactory results can also be achieved by exploiting sky models and known date and location of the scene for outdoor scenarios [156]. Regarding the camera's colors, they can be pre-calibrated using a color chart and a spectroradiometer to measure color values patches in the chart. This process is called colorimetric characterization [143,193]. From these measurements and corresponding values coming from the camera, a 3×3 matrix, $\mathbf{M}$, is computed via least squares optimization. This matrix converts a color from the unknown RGB color space of the camera to XYZ or a known color space. Varghese et al. [390] distilled the previous work and proposed a detailed and rigorous methodology for color calibration.

How many images (N_e)? The minimum required number of images for capturing HDR scenes is two [111]. However, this number strongly depends on the scene, i.e., how HDR values are distributed in it. The naïve methodology that acquires images from the longest to the shortest exposure time (or vice versa) halving (or doubling) shutter time, which is a basically a uniform sampling, is not optimal. In fact, it can produce images sub-optimal SNR [88, 147]. Gallo et al. [138] proposed the first general method for acquiring the optimal number of images in terms of SNR and scene irradiance distribution. Their technique exploits the computational free histograms from the viewfinder preview, which is implemented in most cameras or mobile devices. By collecting histograms at all exposure times, the CDF of the HDR scene histogram is estimated. From this and a camera noise model, the minimum N_e and the shutter times to be set for each image can be computed.

Images alignment and ghosting. The ideal case when capturing HDR images by combining multiple LDR images at different exposure times is to have a fixed camera (i.e., on a tripod) and a static scene (i.e., people/objects/lighting not moving). These ideal conditions occur in controlled environments such as laboratories, a photographic set, etc.

In real world scenarios such ideal conditions are hard to achieve. If an HDR image is merged from a set of LDR images in which either the camera or people/objects move between frames, this will present artifacts in it. Typically, artifacts due to camera movements are called misalignments (creating blurred edges), see Figure 2.6(a) (green box), and artifacts due to people/objects movements are referred to as ghosts (because they partially appear as ghosts in folklore/mythology), see Figure 2.6(a) (red box). Several algorithms for aligning images, removing ghosts, or solving both problems at the same time have been presented. Tursun et al. [375] presented an extensive and exhaustive review of the state-of-the-art techniques for aligning images and removing ghosts when capturing HDR content. These methods can be based on different image processing and computer vision algorithms [361], such as homography alignment through feature detections, optical flow methods, patch synthesis-based methods, Markov random field-based approaches, etc. Amongst all these methods, patch synthesis-based methods are, at the time of writing, the state-of-the-art methods [174, 175, 343]. The concept behind these methods is to synthesize images from ones in the exposure stack by minimizing an energy term such that they should be close to a reference image in stack (the best well-exposed image in the input stack). These methods use PatchMatch methods [53, 160] for accelerating nearest neighbors queries. Although they can generate high quality results (see Figure 2.6(b)) even for complex scenes with large motion, they are computationally expensive;

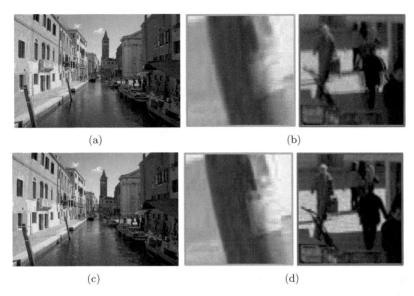

Figure 2.6. An example of artifacts due to scene and camera movements: (a) A tone mapped image from an HDR reconstruction without handling scene (red box) and camera (green box) movements. (b) Zooms of the red and green boxes in (a). (c) A tone mapped image from an HDR reconstruction with the same exposure stack of (a) in input and using a method for removing camera and scene movements [175]. (d) Zooms of the red and green boxes in (c). Note that the image is more sharp (green box) and ghosts free (red box).

e.g., merging an exposure stack of 7 images at 1 Mpixel resolution requires around 3 minutes [343].

Veiling glare. Glare is a phenomenon that reduces visibility because incoming and direct light scatters within the human eye or camera optics [256]. This causes a loss of contrast when acquiring HDR content; see Figure 2.7(a). In order to have glare-free HDR images (see Figure 2.7(b)) Talvala et al. [364] introduced a novel capturing technique in which a mask is placed in front of the camera's lens. This mask blocks glare in some parts of the image. In order to capture all parts of the image glare-free, the mask is moved onto a 6×6 grid in front of the lens, and the scene is captured again in HDR. The final image is generated by merging all these images and removing the estimated glare given the mask pattern. A deconvolution method based on the characterization of the glare for a given camera and lens was also proposed. This was shown to have two main limitations: noise in the dominant glare regions, and the shape of the glare function varies at pixel coordinates making it hard to measure properly.

(a)   (b)

Figure 2.7. An example of veiling glare: (a) A tone mapped HDR image with glare. (b) The tone mapped HDR image in (a) after glare removal.

Note that glare can also be used for encoding HDR information of highlights and bright parts of a scene into LDR images by exploiting cross-screen filters [336]. HDR information can then be recovered using a tomographic reconstruction.

2.1.2 Capturing HDR Videos

The increase in popularity of HDR has driven the move from still images to video content. The first HDR videos had been captured using still images, with techniques such as stop-motion or time-lapse. Under controlled conditions, these methods may provide high quality results with the obvious limitations that stop-motion and time-lapse entail. The method is thus not well suited for real-world situations. More recent methods and technologies for capturing HDR videos are based on multi-sensor systems, spatially varying exposures, multi-view systems, temporally varying exposures, and native sensors.

Multi-sensor systems. Aggarwal and Ahuja [10] proposed one of the first systems where the light path is split into several sensors using one or more beam splitters; see Figure 2.8(a). This makes it possible to capture two or more exposure images at the same time without having ghosts or misalignments at high resolutions and frame rates. To build such systems [86, 367] requires a careful optical alignment, camera calibration, and sensor synchronization. However, small sub-pixel misalignments can still be present.

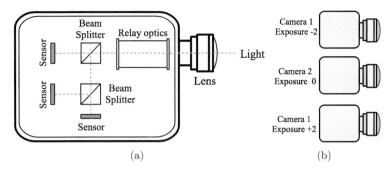

(a)   (b)

Figure 2.8. HDR video acquisition systems: (a) The diagram of a beam splitter setup; three LDR sensors in the same camera have different exposure times in order to capture most of the dynamic range of a scene. (b) The diagram of a multi-view setup; cameras can be arranged either in a linear setup as in the figure or in a grid setup; each camera has a different exposure time.

A solution is to use robust methods for reconstruction [206], which take into account de-Bayering, de-noising, sub-image alignment, and HDR merging into one single solution.

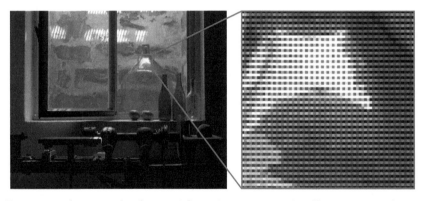

Figure 2.9. An example of a spatial varying exposure time Bayer pattern image. This pattern is not visible when the image is seen at full resolution, which looks like a tone mapped image. A zoom around the neck of the demijohn (red box) elicits the pattern showing the exposure time varying per pixel.

Spatially varying exposure. Nayar et al. [283, 284] proposed varying the exposure spatially, similarly to what occurs with color filters in a Bayer pattern; see Figure 2.9. In this way, there is no need to handle misalignments because all exposure times are captured at the same time. However, spatial resolution is sacrificed for dynamic range, and sophisticated algorithms are required for high quality reconstruction [161, 265, 283, 284, 344].

Researchers have also explored different patterns such as rows varying exposure [91, 157], and non-regular patterns [11]. In this category, Nayar and Branzoi [282] developed an alternative approach, *adaptive dynamic range camera*, where a liquid crystal light modulator is placed in front of the camera. This modulator adapts the exposure of each pixel on the image detector, allowing the capture of scenes with a very large dynamic range.

Multi-view systems. The use of a system with multiple cameras for capturing HDR content [71, 278, 386, 419] provides a solution which does not require an expensive sub-micron sensor alignment [367], and it allows different exposures to be captured at the same time with depth information or lightfields too; see Figure 2.8(b). However, occlusions and specular effects, which both can greatly vary from one view to another, need to be handled with care. These issues become more exacerbated when the baseline is wide. This setup can also be expensive, since more cameras are required and an accurate synchronization system has to be implemented to avoid subtle temporal ghosting.

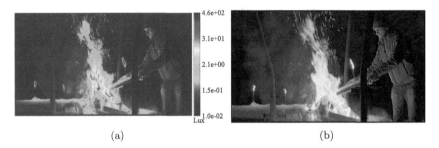

(a)   (b)

Figure 2.10. An example of HDR video capture using a multi-view system with a beam splitter [137]: (a) A frame in false color. (b) The frame in (a) tone mapped. (The original HDR video is courtesy of Jan Fröhlich.)

A hybrid solution [137], in-between this category and multi-sensor systems, is to use multiple cameras and to split the light path in front of the cameras' lenses using a beam splitter (i.e., a semitransparent mirror); see Figure 2.10. This solution solves occlusion and specular effect issues of multi-view systems but may suffer from ghosting artifacts due to double reflections of the transparent mirror at long focal lengths. In this case, a neutral density filter has to be placed in front of the lens of the camera suffering from these artifacts.

Temporally varying exposure. Kang et al. [188] first extended the multiple exposure methods used for images to be used for videos. In this system, a video camera temporally varies its exposure time or ISO at each frame;

see Figure 2.11. As a first step, two adjacent frames are aligned using motion estimation and hierarchical homography. They are then warped and
merged together using a modified version of Equation (2.3). The method
produces high quality results but it cannot handle rapid motion, non-rigid
motions, and significant occlusions. Moreover, ghosting artifacts are the
main drawback of the technique. Researchers have proposed several methods [375] to solve this issue using motion estimation [238], global alignment [12,84], and extending patch-synthesis [187]. Recently, Gryaditskaya
et al. [155] showed that ghosting artifacts can be more disturbing than the
loss of dynamic range. Therefore, they proposed a novel metering algorithm based on the two previous frames' histograms for minimizing such
artifacts during acquisition.

| (a) | (b) | (c) |

Figure 2.11. An example of frames of a temporal varying shutter speed video:
(a) i-th frame (short exposure time). (b) $i + 1$-th (long exposure). (c) Tone
mapped reconstructed HDR frame at time i [12]. (The original video is courtesy
of Tomasz Sergej.)

Native sensors. The alternative to multiple exposure techniques is to use
sensors that can natively capture HDR values [376]. In recent years, sensors
that record into 10/12/14-bit channels in the linear/logarithmic domain
have been introduced by several companies. Relatively low-cost sensors/-
camera systems are now on the market, and they are typically meant for
security and automatization. Main manufacturers are: Cypress Semiconductor [99], Omron [291], PTGrey [313], etc. Then, there are high-quality
cameras with native sensors oriented toward entertainment and cinema,
which provide higher resolutions (e.g., 4K or more) and lower noise levels.
The Arri Alexa 65 [20] and the Sony F65 [350] are two examples of such
cameras.

Thinking outside the box. All methods seen so far are meant for conventional camera sensors (which can be linear or logarithmic) that measure the
intensity of light. Tumblin et al. [371] proposed a camera design for measuring gradients per pixel instead of intensity showing how this can solve
the saturation problem allowing the capture of HDR content. However,
the final HDR image needs a Poisson solver [129] for reconstruction, which

can be computationally expensive. Moreover, this may require substantial changes in existing sensor designs, which may be problematic for camera manufacturers.

Zhao et al. [435] introduced a novel design for capturing HDR content based on a concept similar to phase wrapping of electromagnetic waves. When a sensor pixel saturates, i.e., it reaches the maximum value at n bits, the 2^n modulo is applied to that value; simulating a camera with infinite bits. To have a light and simple design, the number of applications of the modulo are not stored; creating an ill-posed problem during reconstruction. Nevertheless, an original pixel value can be recovered exploiting fringes that are generated in the image by minimizing an energy function via graph cuts [75].

2.1.3 Market-Ready Cameras

HDR cameras. More recently, camera manufactures (e.g., Canon, Nikon, Sony, Sigma, etc.) have introduced the possibility of acquiring 14-bit RAW content increasing the dynamic range captured in one shot. Smartphone and tablet manufactures (e.g., Apple, Samsung, LG, Huawei, etc.) have introduced some HDR capturing features such as automatic exposure bracketing and/or automatic exposure merging. Such features are also offered in standalone cameras. However, dynamic range is limited in the case of smartphones/tablets, i.e., such devices can typically capture only 1–2 extra images; see Figure 2.12, and the final image is stored as a tone mapped LDR image.

(a)   (b)

Figure 2.12. An example of capturing HDR images using a smartphone (an Apple iPhone SE): (a) An image captured with HDR disabled. (b) The same image in (a) with HDR enabled during capture; note that sky details and colors are fully acquired.

HDR videocameras. A number of companies have proposed high quality solutions for the entertainment industry. These are the Thomson Viper camera [149], Arri Alexa cameras [20], the Sony F65 [350], Red Epic cameras [320], the Phantom HD camera [395], and Genesis by Panavision [296]. All these video cameras present high frame rates, low noise, from full HD (1920 × 1080) to 8K resolution (7680 × 4320), and more than 18 f-stops of dynamic range. However, they are extremely expensive and not meant for the consumer market. An overview of an evaluation of the dynamic range of a number of consumer cameras that were tested for imaging rocket launches is given in Karr et al. [190].

Panoramic cameras. A few companies provide 360 panoramic HDR cameras based on automatic multiple exposure capturing that can create radiance maps or tone mapped images. Table 2.1 shows an example of such cameras.

Manufacturer	Device	Dynamic Range (f-stops)	Resolution (pixels)	Time (sec)
LizardQ	LizardQ	30	31800 × 15900	30
Weiss	Civetta	28	14144 × 7072	40
RoundShot	FOVEX Metric	27+	21504 × 10752	145
NCTECH	iSTAR	27	10000 × 5000	10–20
SpheronVR	SpheronLite	26	10400 × 5200	120–420
Panoscan	MK-3*	11	12000 × 6000	54

Table 2.1. A summary of the commercially available HDR spherical cameras. Note: Superscript * refers to data for a single pass, where three passes are needed to obtain an HDR image. Superscript + refers to the fact that the camera captures 15 photographs at different exposure times.

The development of these particular cameras was mainly due to the necessity of quickly capturing HDR images for use in image-based lighting (see Chapter 7), which is extensively used in visual effects, computer graphics, automotive design, and product advertising. Recently, panoramic cameras have received more attention because they can be employed in virtual reality applications. This has sparked the development of many consumer panoramic cameras (e.g., Ricoh Theta, Samsung Gear 360, and Kodak PixPro, to name a few) that are sold at affordable prices (typically less than USD 500). Although they generally cannot capture HDR images natively, there are smartphones/tablets apps that allow users to subsequently compute them.

2.1.4   The Synthesis of HDR Content

Computer graphics rendering methods are another common method of generating HDR content. Frequently, this can be augmented by photographic

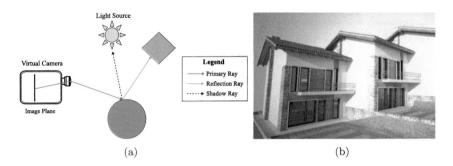

(a)   (b)

Figure 2.13. An example of the-state-of-art of rendering: (a) Ray tracing. For
each pixel in the image, a primary ray is shot through the camera into the scene.
As soon as it hits a primitive, the lighting for the hit point is evaluated. This is
achieved by shooting more rays. For example, a ray toward the light is shot in
the evaluation of lighting. A similar process is repeated for diffuse and specular
inter-reflections. (b) A ray-traced image by Piero Banterle using Maxwell Render
by NextLimit Technologies [286].

methods. Digital image synthesis is the process of rendering images from
virtual scenes composed of formally defined geometric objects, materials,
and lighting, all captured from the perspective of a virtual camera. Two
main algorithms are usually employed for rendering: ray tracing and rasterization.

Ray tracing. Ray tracing [417] models the geometric properties of light
by calculating the interactions of groups of photons, termed *rays*, with
geometry. This technique can reproduce complex visual effects. Rays are
shot from the virtual camera and traverse the scene until the closest object
is hit; see Figure 2.13(a). Here the material properties of the object at that
point are used to calculate the illumination, and a ray is shot toward any
light sources to account for shadow visibility. The material properties at the
intersection point further dictate in which direction reflected/transmitted
rays need to be shot; the process is computed recursively. Due to its
recursive nature, ray tracing and extensions of the basic algorithm, such
as path tracing and distributed ray tracing, are naturally suited to solving
the rendering equation [186], which describes the transport of light within
an environment. Ray tracing methods can thus simulate effects such as
shadows, reflections, refractions, indirect lighting, subsurface scattering,
caustics, motion blur, indirect lighting, and others in a straightforward
manner. While ray tracing is computationally expensive, recent algorithmic
and hardware advances are making it possible to compute it at interactive
rates for dynamic scenes [54, 300].

Rasterization. Rasterization [13] uses a different approach than ray tracing for rendering. The main concept is the projection of each primitive in the scene onto the screen (frame buffer) and, subsequently, its discretization of it into fragments, which are then rasterized into the final image. When a primitive is projected and discretized, visibility has to be solved to have a correct visualization and to avoid incorrect overlap between objects. For this task, the Z-buffer [85] is generally used. The Z-buffer is an image of, typically, the same size as the frame buffer that stores depth values of previous solved fragments. For each fragment at a position $\mathbf{x}$, its depth value, $F(\mathbf{x})_z$, is tested against the stored one in the Z-buffer, $Z(\mathbf{x})_z$. If $F(\mathbf{x})_z < Z(\mathbf{x})_z$, the new fragment is written in the frame buffer, and $F(\mathbf{x})_z$ is placed in the Z-buffer. After the depth test, lighting is evaluated for all fragments. However, shadows, reflections, refractions, and interreflections cannot be handled natively with this process since rays are not shot. These effects are often emulated by rendering the scene from different positions. For example, shadows can be emulated by calculating a Z-buffer from the light source position and applying a depth test during shading to determine if the point is in shadow. This method is known as shadow mapping [420]. The main advantage of rasterization is that it is supported by current graphics hardware, which allows high performances in terms of drawn primitives. Such performance is achieved since it is straightforward to parallelize rasterization: fragments are coherent and independent, and data structures are easy to update. Nevertheless, the emulation of physically based light transport effects (i.e., shadows, reflections/refractions, etc.) is not as accurate as ray tracing and is biased in many cases.

2.1.5 Enhancement and Expansion of LDR Content

The emergence of HDR displays has focused aspects of HDR research onto expanding the many decades of legacy LDR content to take advantage of these new displays. Expansion methods attempt to recreate the missing content for LDR content for which the HDR information was clamped or severely compressed. A number of methods for expanding LDR content have been proposed. Chapter 6 provides a comprehensive overview of these methods. In some cases, LDR content can be enhanced using HDR reference photographs of the same scene. This can be achieved by using techniques [46] similar to chroma-key when the camera is static or by exploiting structure-from-motion for a static scene with a moving camera [61].

2.2 HDR Content Storing

Once HDR content is generated, there is the need to store, distribute, and process these images. An uncompressed HDR pixel is represented using three single precision floating point numbers [173], assuming three bands for RGB colors. This means that a pixel uses 12 bytes of memory, and at a high definition (HD) resolution of 1920×1080 a single image would occupy approximately 24 Mbyte. This is much larger than the approximately 6 Mbytes required to store an equivalent LDR image without compression. Researchers have been working on efficient methods to store HDR content to address the high memory demands. Initially, only compact representations of floating point numbers were used for storing HDR. These methods are still commonly in use in HDR applications and will be covered in this section. More recently, researchers have focused their efforts on compression methods, which will be presented in Chapter 8 and Chapter 9.

HDR values are usually stored using single precision floating point numbers. Integer numbers, which are extensively used in LDR imaging, are not practical for storing HDR values. For example, a 32-bit unsigned integer can represent values in the range $[0, 2^{32} - 1]$, which is not sufficient to cover the entire range experienced by the HVS. It also unsuitable for image processing operations between two or more HDR images; for example, when adding or multiplying, precision can be easily lost and overflow may occur. Such conditions make floating point numbers preferable to integer ones for real-world values [173].

Using single precision floating point numbers, an image occupies 96 bits per pixel (bpp). Ward [405] proposed the first solution to this problem, RGBE, which was originally created for storing HDR values generated by the Radiance rendering system [407]. This method stores a shared exponent between the three colors, assuming that it does not vary much between them. The encoding of the format is defined as

$$E = \lceil \log_2\big(\max(R_w, G_w, B_w)\big) + 128 \rceil,$$

$$R_m = \left\lfloor \frac{256 R_w}{2^{E-128}} \right\rfloor, \quad G_m = \left\lfloor \frac{256 G_w}{2^{E-128}} \right\rfloor, \quad B_m = \left\lfloor \frac{256 B_w}{2^{E-128}} \right\rfloor,$$

and the decoding as

$$R_w = \frac{R_m + 0.5}{256} 2^{E-128}, G_w = \frac{G_m + 0.5}{256} 2^{E-128}, B_w = \frac{B_m + 0.5}{256} 2^{E-128},$$

where $\lfloor \cdot \rfloor$ and $\lceil \cdot \rceil$ are, respectively, the floor and ceiling functions.

Mantissas of the red, R_m, green, G_m, and blue, B_m, channels and the exponent, E, are then each stored in an unsigned char (8-bit), achieving a final format of 32 bpp. The RGBE encoding covers 76 orders of magnitude, but the encoding does not support the full gamut of colors and negative values. To solve this, an image can be converted to the XYZ color space before encoding. This case is referred to as the XYZE format. A modified version of the RGBE format (i.e., 9 bits for mantissa for each color channels, and 5 bits for the shared exponent) is supported by graphics hardware vendors and computer graphics APIs. This format is, respectively, called DXGI_FORMAT_R9G9B9E5_SHAREDEXP [267] and GL_RGB9_E5 in Direct3D and OpenGL [292]. Furthermore, this format can be spatially adapted for efficient image compression [74].

```
function imgRGBE = float2RGBE (img)
%compute shared exponent
[m, n, ~] = size(img);
imgRGBE = zeros(m, n, 4);
v = max(img, [], 3);
[v1, E] = log2(v);
E = E + 128;
v1 = v1 * 256 ./ v;
v1(v < 1e-32) = 0;
E(v < 1e-32) = 0;

%encode R, G, and B
for i=1:3
    imgRGBE(:,:,i) = round(img(:,:,i) .* v1);
end

imgRGBE(:,:,4) = E;
end
```

Listing 2.3. MATLAB code: RGBE encoding [405].

```
function img = RGBE2float (imgRGBE)
[m, n, ~] = size(imgRGBE);
img = zeros(m, n, 3);
%decode E
E = double(imgRGBE(:,:,4) - 128 - 8);
f = pow2(1.0, E);
f(imgRGBE(:,:,4) == 0) = 0;
%decode R, G, and B
for i=1:3
    img(:,:,i) = double(imgRGBE(:,:,i)) .* f;
end
end
```

Listing 2.4. MATLAB code: RGBE decoding [405].

Listing 2.3 and Listing 2.4 show the MATLAB code for encoding and decoding RGBE values from a natively stored HDR image (consisting of a float per color channel).

Ward proposed another HDR image format, a 24/32 bpp perceptually based format, entitled LogLuv [214]. This image format assigns more bits to luminance in the logarithmic domain than to colors in the linear domain. As the first step, an image is converted to XYZ color space and CIE (u', v') chromaticity coordinates are computed (see Appendix B). Then, the 32 bpp format assigns 15 bits to luminance and 16 bits to chromaticity, and it is defined as

$$L_e = \lfloor 256(\log_2 Y_{\mathrm{w}} + 64) \rfloor, \quad u_e = \lfloor 410 u' \rfloor, \quad v_e = \lfloor 410 v' \rfloor, \qquad (2.14)$$

and the decoding as

$$Y = 2^{(L_e + 0.5)/256 - 64}, \quad u' = \left\lfloor \frac{1}{410}(u_e) \right\rfloor, \quad v' = \left\lfloor \frac{1}{410}(v_e) \right\rfloor, \qquad (2.15)$$

where u' and v' are the chromaticity coordinates. The 24 bpp format changes slightly in the constants of Equation (2.14) and Equation (2.15) according to the distribution of bits: 10 bits are allocated to luminance and 14 bits to chromaticity. While the 32-bpp format covers 38 orders of magnitude, the 24-bpp format covers only 4.8 orders of magnitude. The main advantage of LogLuv over RGBE/XYZE is that the format stores luminance and color information separately, making these values directly usable for applications such as tone mapping (see Chapter 3), color manipulation, etc. Furthermore, this format can be improved by exploiting image statistics (i.e., maximum and minimum) [274].

The MATLAB code for LogLuv encoding, based on Equation (2.14), is shown in Listing 2.5. Similarly, the MATLAB code for decoding, using Equation (2.15), is given in Listing 2.6.

```
function imgLogLuv = float2LogLuv(img)
%conversion from RGB to XYZ
imgXYZ = ConvertRGBXYZ(img, 0);

%encode luminance Y
imgLogLuv = zeros(size(img));
Le = floor(256 * (log2(img(:,:,2)) + 64));
imgLogLuv(:,:,1) = ClampImg(Le, 0, 65535);

%CIE (u,v) chromaticity values
norm = (imgXYZ(:,:,1) + imgXYZ(:,:,2) + imgXYZ(:,:,3));
x = imgXYZ(:,:,1) ./ norm;
y = imgXYZ(:,:,2) ./ norm;

%encode chromaticity
norm_uv = (-2 * x + 12 * y + 3);
u_prime = 4 * x ./ norm_uv;
v_prime = 9 * y ./ norm_uv;
```

```
Ue = floor (410 * u_prime );
imgLogLuv (:,:,2) = ClampImg (Ue, 0, 255);

Ve = floor (410 * v_prime );
imgLogLuv (:,:,3) = ClampImg (Ve, 0, 255);
end
```

Listing 2.5. MATLAB code: LogLuv encoding [214].

```
function imgRGB = LogLuv2float (img)
%decode luminance Y
imgXYZ = zeros (size (img));
imgXYZ (:,:,2) = 2.^((img (:,:,1) + 0.5) / 256 - 64);
%decode chromaticity
u_prime = (img (:,:,2) + 0.5) / 410;
v_prime = (img (:,:,3) + 0.5) / 410;

norm = 6 * u_prime -16 * v_prime + 12;
x = 9 * u_prime ./ norm;
y = 4 * v_prime ./ norm;
z = 1 - x - y;

norm = RemoveSpecials (imgXYZ (:,:,2) ./ y);
imgXYZ (:,:,1) = x .* norm;
imgXYZ (:,:,3) = z .* norm;

%conversion from XYZ to RGB
imgRGB = ConvertRGBXYZ (imgXYZ, 1);
end
```

Listing 2.6. MATLAB code: LogLuv decoding [214].

Another common HDR image format is the half-floating point format, which is part of the specification of the OpenEXR file format [180]. In this representation, each color is encoded using a half precision floating point number, which is a 16-bit implementation of the IEEE 754 standard [173], and it is defined as

$$
H = \begin{cases}
0 & \text{if } \left(M = 0 \wedge E = 0\right), \\
(-1)^S 2^{E-15} + \frac{M}{1024} & \text{if } E = 0, \\
(-1)^S 2^{E-15}\left(1 + \frac{M}{1024}\right) & \text{if } 1 \leq E \leq 30, \\
(-1)^S \infty & \text{if } \left(E = 31 \wedge M = 0\right), \\
\text{NaN} & \text{if } \left(E = 31 \wedge M > 0\right),
\end{cases}
$$

where S represents the sign, occupying 1 bit, M is the mantissa, occupying 10 bits, and E is the exponent, occupying 5 bits. Therefore, the final format is 48 bpp, covering around 10.7 orders of magnitude. The main advantage, despite the size, is that this format is implemented in graphics hardware allowing real-time applications to use HDR images. This format is considered the *de facto* standard in the movie industry [112].

Several medium dynamic range formats, which have the purpose of covering the classic film range between 2–4 orders of magnitude, have been proposed by the entertainment industry. However, they are not suitable for HDR images/videos. The log encoding image format created by Pixar is one such example [112].

2.3   Native Visualization

Display technologies that can natively visualize HDR images and videos without using TMOs are now becoming available. The first such device was the HDR viewer [218, 408], which was inspired by the classic stereoscope; a device used at the turn of the nineteenth to twentieth centuries for displaying 3D images. Although this device can achieve a dynamic range of 10,000:1, it is very limited. In fact, it can display only printed static HDR still images with a high cost for printing, i.e., USD 50. The HDR monitor [342] was the first display to visualize HDR content. Table 2.2 shows an overview on these early devices. The methods to display content on these devices both divide an HDR image into a detail layer with colors and a luminance layer that back-modulates the first layer.

Device	White Point (cd/m^2)	Black Point (cd/m^2)	Dynamic Range
HDR Monitor 14"(Projector) [342]	2,700	0.054	50,000:1
HDR Monitor 18"(LED) [342]	8,500	0.03	283,333:1
HDR Monitor 37"(LED) [76]	3,000	0.015	200,000:1

Table 2.2. The features of early HDR display devices.

2.3.1   HDR Monitors

Seetzen et al. [342] developed the first HDR monitors. These were based on two technologies: a digital light processing (DLP) projector, and light emitting diodes (LEDs). There is a modulated light source that boosts the dynamic range of a front layer that encodes details and colors. Both the DLP and LEDs HDR monitors use LCDs for displaying the front layer.

The DLP projector driven HDR display was the first of these technologies to be developed. This system uses a DLP projector to modulate the light; see Figure 2.14. In order to display HDR content on this device, an/a image/frame needs to be preprocessed to obtain an input for the light

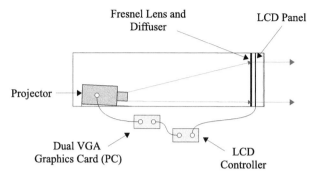

Figure 2.14. A scheme of the projector-based HDR display.

modulator (i.e., the DLP projector in this case) and the front panel (i.e., a LCD monitor in this case). The processing method is shown in Figure 2.15, and it creates two images. The first one, which modulates the light source, is created by subsequently applying the square root and projector inverse response to the luminance of the original image. The second image, placed in front of the one for modulation, is generated by dividing the HDR image by the modulation one (after taking into account the PSF of the projector) and applying the inverse response of the LCD panel. Note that while the image in front encodes colors and details, the back one is used for modulation and encodes the global luminance distribution.

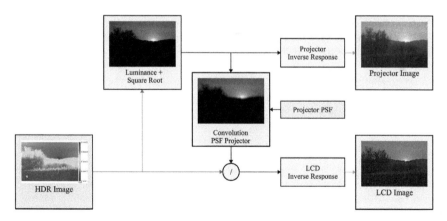

Figure 2.15. The processing pipeline to generate two images for the projector-based HDR monitor by Seetzen et al. [342].

The measured dynamic range of the 15.1" projector driven prototype monitor is 50,000:1, where the measured maximum luminance level is 2,700

cd/m^2 and the minimum luminance level is 0.054 cd/m^2. For this monitor, the LCD panel is a 15.1" Sharp LQ150X1DG0 with a dynamic range of 300:1, and the projector is an Optoma DLP EzPro737 with a contrast ratio of 800:1. However, this technology is impractical under most conditions. The required optical path for the projector is large, around 1m for a 15.1" monitor, which is not practical for home entertainment and wider displays. Moreover, the viewing angle is very small because a Fresnel lens, which is used for uniforming luminance values, has a huge fall-off at wide viewing angles. Finally, the projector needs to be very bright, and this entails high power consumption and significant heat generation.

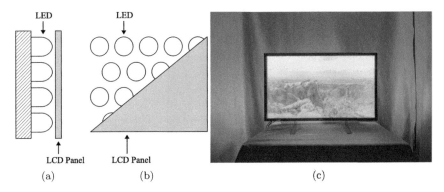

Figure 2.16. The HDR monitor based on LCD and LED technologies: (a) The scheme of a part of the monitor in a lateral section (b) The scheme of a part of the monitor in a frontal section (c) The first commercial HDR display, the SIM2 Grand Cinema^{TM} SOLAR 47. (The photograph was taken with the help of Giuseppe Valenzise and Frédéric Dufaux.)

The LED-based technology uses a low resolution LED panel to modulate the light; see Figure 2.16(a) and Figure 2.16(b). The processing algorithm for the generation of images for the LED and the LCD panels is similar to the one for the DLP device. The main difference is the addition of a step in which the luminance for each LED is determined based on a downsampled square root luminance to the resolution of the LED panel and iteratively solving for the values, taking the overlap of the point spread function of the LED into account [369].

The main LED model is the DR-37p, a 37" HDR monitor with a 200,000:1 dynamic range where the measured maximum and minimum luminance level are, respectively, 3,000 cd/m^2 and 0.015 cd/m^2. For this monitor, 1,380 Seoul Semiconductor PN-W10290 LEDs were mounted behind a 37" Chi Mei Optoelectronics V370H1L01 LCD panel with a contrast ratio of 250:1 [76].

This display technology did have some issues to solve before becoming a fully fledged consumer product. The first one is the quality. While the dynamic range is increased, the image quality is reduced having a lower back-modulated resolution than the projector technology. The second one is the LEDs which, at the time the device came out, were very expensive, consumed a lot of power (1,680 W), and required cooling. The heat dissipation is carried out using fans and a liquid-based system, but this results in quite a lot of noise. Wetzstein et al. [416] introduced a tomographic method for tackling the visualization of lightfields using a multi-layer display. This method can also be used for displaying HDR content. They showed that their tomographic solver and dual-layer decomposition [342] are qualitatively-similar; indicating that dual-layer HDR displays are near-optimal. Nevertheless, their solution can increase the dynamic range but not equally well for all spatial frequencies. This is an important result, which is valid for dual-layer HDR displays as well, because it shows that the optimal performance can happen only without layer separation; although this can be difficult to achieve in practice.

In February 2007, Dolby acquired the HDR display pioneers Brightside producers of the 15.1" projector driven prototype monitor and DR-37p, discussed above. This HDR technology was licensed to the Italian company SIM2 that announced the first commercial HDR display, the Solar 47, in 2009; see Figure 2.16(c). The latest commercial iteration of the SIM2 HDR monitor is the HDR47ES6MB, which has a white peak of over 6,000 cd/m^2 at 1920×1080 pixels resolution with 2202 LEDs in the backlit panel. At the IBC event in Amsterdam in September 2016, SIM2 showed a prototype 10,000 cd/m^2 display. Given the high luminance output, these devices are suitable for research, high quality home experience, and industrial visualization.

Note that the colorometrical calibration of dual-layer HDR displays [342] poses more challenges than traditional displays for colorimetric characterization because these devices potentially have a 4D space to be calibrated; i.e., LEDs (or a projector) and an LCD. Nevertheless, a general solution to display calibrated content from a 3D color space can be derived by using look-up tables and affine transformations [338]. Recently, Liu et al. [230] showed that the use of thin-plate splines and 3D octree forests can reduce the error during color reproduction based on samples covering the whole output gamut volume.

2.3.2 HDR Printing

Even though smartphones and tablets have become one of the major forms of display media in our society, printed media still has an important role to play. For example, medical doctors still prefer x-ray films for examination, and photography printouts have an important role for museums and for furnishing homes. Therefore, delivering an HDR experience using classic media such as paper prints is important.

One of the first attempts is to enhance a paper print (or an e-paper display) by *superimposing* it onto a projected image [63]. As a first step, the printout and the projected image need to be geometrically registered via a camera [361]. Then, the camera-projector system is photometrically calibrated [77] for linearization; this applies also to the printer's transfer function. Finally, the image to be printed, I_A, and the image to be projected, I_B, are computed using a splitting technique similar to Seetzen et al. [342].

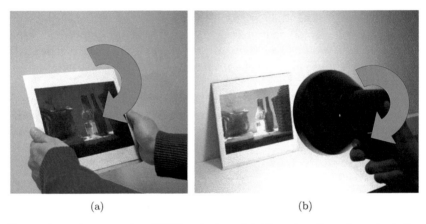

(a)   (b)

Figure 2.17. An example of HDR printing: (a) Changing the print orientation. (b) Changing the print lighting.

In order to obtain a lighter and more unconstrained setup than HDR superimposition, Dong et al. [114] proposed to print HDR images using different glossy inks at different dynamic range areas of the image to be printed. Note that glossy inks can vary their appearance at different lighting and viewing conditions; spanning their appearance from dark to highlights. This allows users to have an HDR experience either by changing the print orientation or by changing its lighting conditions; see Figure 2.17. As a first step, they characterized the BRDFs of different glossy inks. Then, given an input HDR image, the algorithm determines the ink combination of all pixels and the exposure settings for all viewing conditions. This is achieved by minimizing via least squares optimization the differences between the printed HDR image appearance (for a given lighting condition or view) and the appearance of the input HDR image at the corresponding exposure. Although the setup is not cumbersome, i.e., an external light source, it requires the viewing environment does not have other strong light sources.

2.3.3 Summary

The HDR pipeline has recently reached maturity. There are several methods for acquiring both HDR still images and HDR videos at different costs, which are also available to consumers at a satisfactory quality. This content can now also be efficiently stored and streamed around the web; see Chapter 8 and Chapter 9 for more information on this topic. Although tone mapping has been the main way to experience HDR content, see Chapter 3 and Chapter 5 for an overview, native visualization of HDR content is quickly appearing in the market and will probably become the dominant method for HDR viewing over the next decade.

3
Tone Mapping

Most display devices available nowadays are not able to natively display HDR content. Entry level displays have a low contrast ratio of only around 200:1. Although high-end LCD televisions have a much higher contrast ratio, on average around 10,000:1, they are typically discretized at 8-bit and rarely at 10-bit per color channel. This means that color shades are limited to 255. In the last two decades, researchers have spent significant time and effort in order to compress the range of HDR images and videos so the data may be visualized more "naturally" on LDR displays.

Tone mapping is the operation that adapts the dynamic range of HDR content to suit the lower dynamic range available on a given display. This reduction of the range attempts to keep some characteristics of the original

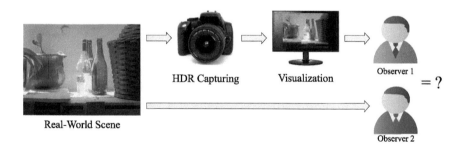

Figure 3.1. The relationship between tone mapped and real-world scenes. Observer 1 and Observer 2 are looking at the same scene but in two different environments. Observer 1 is viewing the scene on a monitor after it has been captured, stored, and tone mapped. Observer 2, on the other hand, is watching the scene in the real world. The final goal is that the tone mapped scene should match the perception of the real-world scene and thus Observers 1 and 2 will perceive the same scene.

content such as local and global contrast, details, etc. Furthermore, it is
often desired that the perception of the tone mapped image should match
the perception of the real-world scene; see Figure 3.1. Tone mapping is
performed using an operator f or tone mapping operator (TMO), which is
defined in general as

$$f(I) : \mathbb{R}_i^{w \times h \times c} \to \mathbb{D}_o^{w \times h \times c}, \tag{3.1}$$

where I is the image, and w, h, and c are, respectively, the width, the
height, and the number of color bands of I. $\mathbb{R}_i \subseteq \mathbb{R}$, $\mathbb{D}_o \subset \mathbb{R}_i$, where
$\mathbb{D}_o = [0, 255]$ for normal LDR monitors. Furthermore, only luminance
is usually tone mapped by a TMO, while colors are unprocessed. This
simplifies Equation (3.1) to

$$f(I) = \begin{cases} L_{\mathrm{d}} = f_{\mathrm{L}}(L_{\mathrm{w}}) : \mathbb{R}_i^{w \times h} \to [0, 255], \\ \begin{bmatrix} R_{\mathrm{d}} \\ G_{\mathrm{d}} \\ B_{\mathrm{d}} \end{bmatrix} = L_{\mathrm{d}} \left(\frac{1}{L_{\mathrm{w}}} \begin{bmatrix} R_{\mathrm{w}} \\ G_{\mathrm{w}} \\ B_{\mathrm{w}} \end{bmatrix} \right)^s, \end{cases} \tag{3.2}$$

where $s \in (0, 1]$ is a saturation factor that decreases saturation. This
is usually increased during tone mapping. After the application of f,
gamma correction is usually applied and each color channel is clamped
to the range $[0, 255]$.

Note that the original gamut is greatly modified in this process, and the
tone mapped color appearance can result in great differences from that in
the original image. Research addressing this issue is presented in Chapter 4.

TMOs can be broadly classified into different groups based on f or the
image processing techniques they use; see Table 3.1.

	Global	Local	Frequency	Segmentation
Empirical	LM [341], ELM [341], QT [341]	SVTR [90], PTR [323]	LICS [374], BF [118], GDC [129], CHI [225]	IM [229], EF [261]
Perceptual	PBR^{T} [372], CBSF [406], VAM [130], HA [215], TDVA^{T} [302], AL [116]	TMOHCI [27], LMEA^{T} [217]		SA [428], LP [204]

Table 3.1. The taxonomy of TMOs presented in this chapter. TMOs are divided
based on their image processing techniques. ^{T} means that this operator may be
suitable for tone mapping HDR videos. See Table 3.2 for a clarification of the
key.

The main groups of the taxonomy are

Key	Name
AL	Adaptive Logarithmic
BF	Bilateral Filtering
CHI	Companding HDR Images
CSBF	Contrast Based Scale Factor
EF	Exposure Fusion
ELM	Exponential/Logarithmic Mapping
GDC	Gradient Domain Compression
HA	Histogram Adjustment
IM	Interactive Manipulation
LCSI	Low Curvature Image Simplifiers
LM	Linear Mapping
LMEA	Local Model of Eye Adaptation
LP	Lightness Perception
PBR	Perceptual Brightness Reproduction
PTR	Photographic Tone Reproduction
QT	Quantization Technique
SA	Segmentation Approach
SVTR	Spatially Variant Tone Reproduction
TDVA	Time Dependent Visual Adaptation
TMOHCI	Tone Mapping Operator for High Contrast Images
VAM	Visual Adaptation Model

Table 3.2. Key to TMOs for Table 3.1.

- Global operators. The mapping is applied to all pixels with the same operator f.

- Local operators. The mapping of a pixel depends on its neighbors, which are given as an input to f.

- Segmentation operators. The image is segmented in broad regions, and a different mapping is applied to each region.

- Frequency/gradient operators. The lower and higher frequencies of the images are separated. While an operator is applied to the lower frequencies, the higher frequencies are usually maintained as they preserve fine details.

Further classifications can be given based on the design philosophy of the TMO, or its use:

- Perceptual operators. These operators can be global, local, segmentation, or frequency/gradient. The main focus is that the function f models some aspects of the HVS, such as temporal adaptation [302], loss of visual acuity at night [130], aging [182], etc.

- Empirical operators. These operators can be global, local, segmentation, or frequency/gradient. In this case, f does not try to mimic the

HVS, but attempts to generate aesthetically pleasing images inspired by other fields, such as photography [323].

Finally, certain methods are suitable for HDR video content and animations, while others are typically only used for images. The HDR video operators are discussed in Chapter 5.

In the next section, the MATLAB framework is presented and some basic common functions are described. Subsequently, the main TMOs are reviewed. This review is organized by image processing techniques described by the taxonomy in Table 3.1; this table also shows the operators introduced in this chapter and their position in the taxonoomy. Global operators are discussed in Section 3.2 local operators in Section 3.3, frequency operators in Section 3.4, and segmentation operators in Section 3.5.

3.1 TMO MATLAB® Framework

Often TMOs, independently to which category they belong, have two common steps. This section describes the common routines that are used by most, but not all, TMOs. The first step is the extraction of the luminance information from the input HDR image or frame. This is because a TMO is typically working on the luminance channel, avoiding color compression. The second step is the restoration of color information in the compressed image. The implementation of these two steps is shown in Listing 3.1 and Listing 3.2.

In the first step of Listing 3.1, the input image, `img`, is checked to see if it is composed of three color channels. Then the luminance channel is extracted using the function `lum.m`, under the folder `ColorSpace`. Note that for each TMO that will be presented in this chapter, optional input parameters for determining the appearance of the output image, `imgOut`,

```
%is it a gray-scale or an three color channels image?
check13Color(img);

%does img have negative values?
checkNegative(img);

%compute luminance
L = lum(img);
```

Listing 3.1. MATLAB code: TMO first step.

are verified. If not present, they are set equal to default values suggested by their authors.

In the last step of Listing 3.2, `ChangeLuminance.m` in the folder `Tmo/util` is applied to `img` to remove the old luminance, L, and substituting it with `Ld`, obtaining `imgOut`.

An optional step is color correction. Many TMOs handle this by applying Equation (3.2) to the final output. However, this extra process is left out because color appearance can substantially vary depending on the TMO's parameters. This function, `ColorCorrection.m`, under the folder `ColorCorrection`, is shown in Listing 3.3, and applies Equation (3.2) to the input image in a straightforward way. Note that the correction value, `cc_s`, can be a single channel image per pixel correction.

```
imgOut = ChangeLuminance(img, L, Ld);
```

Listing 3.2. MATLAB code: TMO last step.

```
L = lum(img);
imgOut = zeros(size(img));

for i=1:size(img, 3);
    imgOut(:,:,i) = ((img(:,:,i) ./ L).^cc_s) .* L;
end

imgOut = RemoveSpecials(imgOut);
```

Listing 3.3. MATLAB code: color correction (i.e., desaturation/saturation) for images (`ColorCorrection.m`).

All implemented TMOs in this book produce linearly tone mapped values in $[0, 1]$. In order to display tone mapped images properly on a display, the inverse characteristic of the monitor needs to be applied. A straightforward way to do this for standard LCD and CRT monitors is to apply an inverse gamma function, typically with $\gamma = 2.2$, by calling `GammaTMO.m` function in the folder `Tmo`, or applying $sRGB$ non-linear function by calling the `ConvertRGBtosRGB.m` function in the folder `ColorSpace`.

3.2 Global Operators

With global operators, the same operator f is applied to all pixels of the input image, preserving global contrast. The operator may sometimes perform a first pass of the image to calculate image statistics; these are subsequently used to optimize the dynamic range reduction. Some common statistics that are typically calculated for tone mapping are maximum luminance, minimum luminance, and logarithmic or arithmetic average values;

see Section 1.1.3. To increase robustness and to avoid outliers, these statistics are calculated using percentiles, especially for minimum and maximum values, because they could have been affected by noise during image capture. It is relatively straightforward to extend global operators into the temporal domain. In most cases it is sufficient to temporally filter the computed image statistics, thus reducing or avoiding possible flickering artifacts due to the temporal discontinuities of the frames in the sequence. The main drawback of global operators is that, since they make use of global image statistics, they are unable to maintain local contrast and the finer details of the original HDR image.

3.2.1 Straightforward Mapping Methods

Straightforward operators are based on practical functions, such as linear scaling, logarithmic functions, and exponential functions. While they are usually fast and simple to implement, they cannot fully compress the dynamic range accurately.

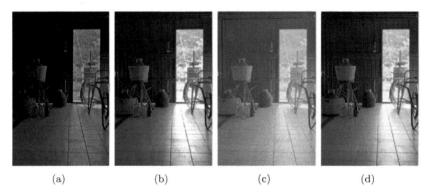

(a)   (b)   (c)   (d)

Figure 3.2. An example of the applications of simple operators to the Cellar HDR image: (a) Normalization. (b) Automatic exposure. (c) Logarithmic mapping $q = 0.0$ and $k = 1$. (d) Exponential mapping $k = 0.1$.

Linear exposure is a straightforward way to visualize HDR images. The starting image is multiplied by a factor e that is similar in concept to the exposure used in a digital camera:

$$L_d(\mathbf{x}) = e \cdot L_w(\mathbf{x}).$$

The user chooses e based on information that is considered interesting to visualize. When $e = \frac{1}{L_{w,\,max}}$, this scaling is called normalization and it can result in a very dark appearance; see Figure 3.2(a). If e is calculated by maximizing the number of well-exposed pixels, the scaling is called automatic exposure [285]; see Figure 3.2(b). Ward [406] proposed to calculate *e* by matching the threshold visibility in an image and a display using TVI functions; see Section 1.1.2. However, a simple linear scale cannot compress the dynamic range of the scene, hence it shows only a slice of information.

Logarithmic mapping applies a logarithm function to HDR values. The base is the maximum value of the HDR image to map nonlinear values in the range $[0, 1]$. The operator is defined as

$$L_d(\mathbf{x}) = \frac{\log_{10}\big(1 + qL_w(\mathbf{x})\big)}{\log_{10}\big(1 + kL_{w,\,max}\big)}, \tag{3.3}$$

where $q \in [1, \infty)$ and $k \in [1, \infty)$ are constants selected by the user for determining the desired appearance of the image.

```
LMax = max(L(:)); %compute the max luminance

%dynamic range reduction
Ld = log10(1 + L * log_q) / log10(1 + LMax * log_k);
```

Listing 3.4. MATLAB code: Logarithmic TMO (`LogarithmicTMO.m`).

Listing 3.4 provides the MATLAB code of the logarithm mapping operator. The full code can be found in the file `LogarithmicTMO.m`. The variables `log_q` and `log_k` are respectively equivalent to the parameter q and k in Equation (3.3). Figure 3.2(c) shows an example of an image tone mapped using the logarithm mapping operator for $q = 0.0$ and $k = 1.0$.

Exponential mapping applies an exponential function to HDR values. It maps values to the interval $[0, 1]$, where each value is divided by the arithmetic average. The operator is defined as

$$L_d(\mathbf{x}) = 1 - \exp\left(-k\frac{L_w(\mathbf{x})}{L_{w,\,H}}\right), \tag{3.4}$$

where $k \in (0, \infty)$ is a parameter selected by the user.

```
Lwa = logMean(L); %luminance geometric mean

%dynamic range reduction
Ld = 1 - exp(-exp_k * ( L / Lwa));
```

Listing 3.5. MATLAB code: Exponential TMO (`ExponentialTMO.m`).

Listing 3.5 provides the MATLAB code of the exponential mapping technique. The full code can be found in the file `ExponetialTMO.m`. The variable `exp_k` is k in Equation (3.4). Figure 3.2(d) shows the use of the exponential mapping operator for $k = 0.1$.

Both exponential and logarithmic mapping can deal with medium dynamic range content reasonably well. However, these operators struggle when attempting to compress full HDR content. This can result in a very dark or bright appearance of the image, low preservation of global contrast, and an unnatural look.

3.2.2   Brightness Reproduction

One of the first TMOs in the field of computer graphics was proposed by Tumblin and Rushmeier [373]; it was subsequently revised by Tumblin et al. [372]. This operator is inspired by the HVS and adopts Stevens and Stevens' work on brightness [353, 354]. The TMO is defined as

$$L_\mathrm{d}(\mathbf{x}) = \frac{1}{L_\mathrm{d,\,max}} \cdot m \cdot L_\mathrm{d,\,a} \left(\frac{L_\mathrm{w}(\mathbf{x})}{L_\mathrm{w,\,H}} \right)^\alpha, \qquad \alpha = \frac{\gamma(L_\mathrm{w,\,H})}{\gamma(L_\mathrm{d,\,a})}, \qquad (3.5)$$

where $L_\mathrm{d,\,a}$ is the adaptation luminance of the display (30–100 cd/m^2 for LDR displays). The function $\gamma(x)$ is the Stevens and Stevens' CSF for a human adapted to a luminance value x, which is defined as

$$\gamma(x) = \begin{cases} 1.855 + 0.4 \log_{10}(x + 2.3 \cdot 10^{-5}) & \text{for } x \leq 100 \text{ cd/m}^2, \\ 2.655 & \text{otherwise.} \end{cases} \qquad (3.6)$$

Finally, m is the adaptation-dependent scaling term, which prevents anomalous gray night images. It is defined as

$$m = \left(C_\mathrm{max} \right)^{\frac{\gamma_\mathrm{wd} - 1}{2}}, \qquad \gamma_\mathrm{wd} = \frac{\gamma(L_\mathrm{w,\,H})}{1.855 + 0.4 \log_{10}(L_\mathrm{d,\,a})}, \qquad (3.7)$$

where C_max is the maximum contrast of the display.

This operator compresses HDR images, preserving brightness and producing plausible results when calibrated luminance values are available. Note that this TMO needs gamma correction to avoid dark images even if it already includes a power function; see the parameter α in Equation (3.5). Figure 3.3 shows some results.

Listing 3.6 provides the MATLAB code of Tumblin et al. operator [373]. The full code can be found in the file TumblinTMO.m. The method takes the following parameters of the display as input: the adaptation luminance of the display, L_da (default is 20 cd/m^2), the maximum luminance of the display, Ld_Max (default is 100 cd/m^2), and maximum contrast, C_Max (default is 100). Lwa is $L_\mathrm{w,\,H}$ in Equation (3.5) and Equation (3.7), and it is computed as the geometric mean of img using logMean.m in the folder Tmo/util.

The next step is to compute the Stevens and Stevens' CSF $\gamma(x)$ for the luminance adaptation of the two observers (real-world input image and displayed image). This is calculated using the function StevensCSF.m, which implements Equation (3.6), and it can be found in the Tmo/util folder; see Listing 3.7.

(a)   (b)

Figure 3.3. An example of Tumblin et al. [373] applied to an HDR image for an
LDR monitor (with $C_{\max} = 100$, and $L_{d,\ \max} = 250\mathrm{cd/m^2}$): (a) A tone mapped
image with $L_{d,\ a} = 30\ \mathrm{cd/m^2}$. (b) A tone mapped image with $L_{d,\ a} = 100\ \mathrm{cd/m^2}$.
Note that the overall image brightness is directly proportional to $L_{d,\ a}$.

```
L_wa = logMean(L); %luminance world adaptation

%range compression
gamma_w = StevensCSF(L_wa);
gamma_d = StevensCSF(L_da);
gamma_wd = gamma_w / (1.855 + 0.4 * log(L_da));
m = C_Max.^((gamma_wd - 1) / 2.0);
Ld = L_da * m .* ((L / L_wa).^(gamma_w / gamma_d));
Ld = Ld / Ld_Max;
```

Listing 3.6. MATLAB code: Tumblin et al. operator [373] (TumblinTMO.m).

Afterward, the adaptation-dependent scaling term m is calculated as in
Equation (3.7), which corresponds to m in the code. Finally, the compressed
luminance Ld is computed using Equation (3.5).

```
function y = StevensCSF(x)

y = zeros(size(x));
y(x <= 100) = 1.855 + 0.4 * log10(x(x <= 100) + 2.3 * 1e-5);
y(x > 100) = 2.655;

end
```

Listing 3.7. MATLAB code: Stevens and Stevens' CSF (StevensCSF.m).

3.2.3   Quantization Techniques

Schlick [341] proposed an operator based on rational functions for providing a straightforward and intuitive approach to tone mapping. The TMO is defined as

$$L_d(\mathbf{x}) = \frac{pL_w(\mathbf{x})}{(p-1)L_w(\mathbf{x}) + L_{w,\,max}}, \qquad (3.8)$$

where $p \in [1, \infty)$ and it can be automatically estimated as

$$p = \frac{L_0 L_{w,\,max}}{2^N L_{w,\,min}}. \qquad (3.9)$$

The variable N is the number of bits of the output display, and L_0 is the lowest luminance value of a monitor that can be perceived by the HVS. The use of p in Equation (3.9) is a uniform quantization process since the same function is applied to all pixels. A non-uniform quantization process can be adopted using an adaptation parameter p that is spatially varying (p') computed as

$$p' = p\left(1 - k + k\frac{L_{w,\,avg}(\mathbf{x})}{\sqrt{L_{w,\,max}L_{w,\,min}}}\right), \qquad (3.10)$$

where $k \in [0, 1]$ is a weight of non-uniformity that is chosen by the user, and $L_{w,\,avg}(\mathbf{x})$ is the average luminance intensity of a given zone surrounding the pixel. The behavior of this non-uniform process is commonly associated with a local operator. Schlick used $k = 0.5$ in all his experiments. Three different techniques to compute the average luminance intensity value $L_{w,\,avg}(\mathbf{x})$ were also proposed; for more details refer to Schlick's original article [341]. This non-uniform process is justified by the fact that

(a)	(b)

Figure 3.4. An example of quantization techniques applied to the Squero HDR image: (a) The uniform technique using automatic estimation for p; see Equation (3.9). (b) The non-uniform technique with $k = 1$.

the human eye moves continuously from one point to another in an image. For each point on which the eye focuses there exists a surrounding zone that creates a local adaptation and modifies the luminance perception. The quantization technique provides a simple and computationally fast TMO; see Figure 3.4. However, a user needs to specify an appropriate k value for each image.

```
%compute max luminance
LMax = MaxQuart(L, 0.99);
%comput min luminance
LMin = MaxQuart(L(L > 0.0), 0.01);

%mode selection
switch s_mode
    case 'manual'
        p = max([p, 1]);

    case 'automatic'
        p = L0 * LMax / (2^nBit * LMin);

    case 'nonuniform'
        p = L0 * LMax / (2^nBit * LMin);
        p = p * (1 - k + k * L / sqrt(LMax * LMin));
end

%dynamic range reduction
Ld = p .* L ./ ((p - 1) .* L + LMax);
```

Listing 3.8. MATLAB code: Schlick TMO [341].

Listing 3.8 provides the MATLAB code of the Schlick TMO [341]. The full code may be found in the file SchlickTMO.m. The parameter s_mode specifies the type of model of the Schlick technique used. There are three modes: manual, automatic, and nonuniform. The manual mode takes the parameter p as input from the user. The automatic and nonuniform modes use the uniform and non-uniform quantization technique, respectively. The variable p is either p or p' depending on the mode used, nBit is the number of bits N of the output display, L0 is L_0, and k is k. The first step is to compute the maximum, L_Max, and the minimum luminance, L_Min in a robust way (to avoid unpleasant appearances), i.e., computing percentiles. These values can be used for calculating p. Afterward, based on s_mode, one of the three modalities is chosen and either p is given by the user (manual mode) or is computed using Equation (3.9) (automatic mode) or using Equation (3.10) (nonuniform mode). Finally, the dynamic range of the luminance channel is reduced by applying Equation (3.8).

3.2.4   A Model of Visual Adaptation

Ferwerda et al. [130] presented a TMO based on subjective experiments, which was subsequently extended by Durand and Dorsey [117]. This operator models aspects of the HVS such as changes in threshold visibility, color appearance, visual acuity, and sensitivity over time, which depend on adaptation mechanisms of the HVS; see Figure 3.5.

(a)   (b)   (c)

Figure 3.5. An example of the operator proposed by Ferwerda et al. [130], varying the mean luminance of the input HDR image: (a) 1 cd/m^2. (b) 10 cd/m^2. (c) 100 cd/m^2. Note that colors vanish when decreasing the mean luminance due to the nature of scotopic vision.

This is achieved by using TVI functions for modeling cones (TVI$_p$, photopic vision) and rods (TVI$_s$, scotopic vision). These functions are defined as

$$\log_{10} \text{TVI}_p(x) = \begin{cases} -0.72 & \text{if } \log_{10} x \leq -2.6, \\ \log_{10} x - 1.255 & \text{if } \log_{10} x \geq 1.9, \\ (0.249 \log_{10} x + 0.65)^{2.7} - 0.72 & \text{otherwise,} \end{cases}$$

and

$$\log_{10} \text{TVI}_s(x) = \begin{cases} -2.86 & \text{if } \log_{10} x \leq -3.94, \\ \log_{10} x - 0.395 & \text{if } \log_{10} x \geq -1.44, \\ (0.405 \log_{10} x + 1.6)^{2.18} - 2.86 & \text{otherwise.} \end{cases}$$

The operator is a linear scale of each color channel for simulating photopic conditions, added to an achromatic term for simulating scotopic conditions.

The operator is given by

$$
\begin{bmatrix} R_\mathrm{d}(\mathbf{x}) \\ G_\mathrm{d}(\mathbf{x}) \\ B_\mathrm{d}(\mathbf{x}) \end{bmatrix} = m_c(L_\mathrm{d,\,a}, L_\mathrm{wa}) \begin{bmatrix} R_\mathrm{w}(\mathbf{x}) \\ G_\mathrm{w}(\mathbf{x}) \\ B_\mathrm{w}(\mathbf{x}) \end{bmatrix} + m_r(L_\mathrm{d,\,a}, L_\mathrm{wa}) \begin{bmatrix} L_\mathrm{w}(\mathbf{x}) \\ L_\mathrm{w}(\mathbf{x}) \\ L_\mathrm{w}(\mathbf{x}) \end{bmatrix}, \quad (3.11)
$$

where $L_\mathrm{d,\,a}$ is the luminance adaptation of the display (e.g., $L_\mathrm{d,\,a} = L_\mathrm{d,\,max}/2$, where $L_\mathrm{d,\,max}$ is the maximum luminance level of the display), L_wa is the luminance adaptation of the image (e.g., $L_\mathrm{wa} = L_\mathrm{w,\,max}$), and m_r and m_c are two scaling factors that depend on the TVI functions. They are defined as

$$
m_c(L_\mathrm{d,\,a}, L_\mathrm{wa}) = \frac{\mathrm{TVI}_p(L_\mathrm{d,\,a})}{\mathrm{TVI}_p(L_\mathrm{wa})}, \qquad m_r(L_\mathrm{d,\,a}, L_\mathrm{wa}) = \frac{\mathrm{TVI}_s(L_\mathrm{d,\,a})}{\mathrm{TVI}_s(L_\mathrm{wa})}.
$$

Durand and Dorsey [117] extended this operator to work in the mesopic vision based on work of Walraven and Valeton [397]. They also added a time-dependent mechanism using data from Adelson [5] and Hayhoe [168]. Note that a more accurate simulation of scotopic conditions can be achieved by taking into account loss of acuity [366], scotopic noise [191], and accurate simulation of the Purkinje effect [195] (the blue color shift in low light conditions).

The TMO simulates many aspects of the HVS, but the reduction of the dynamic range is achieved through a simple linear scale that cannot apply drastic range compression.

```
%compute the scaling factors
mC = TpFerwerda(L_da) / TpFerwerda(L_wa); %cones
mR = TsFerwerda(L_da) / TsFerwerda(L_wa); %rods
k = WalravenValeton_k(L_wa);

%scale the HDR image
col = size(img,3);
imgOut = zeros(size(img));

switch col
    case 3
        vec = [1.05, 0.97, 1.27];
    otherwise
        vec = ones(col, 1);
end

for i=1:col
    imgOut(:,:,i) = (mC * img(:,:,i) + vec(i) * mR * k * L);
end

imgOut = ClampImg(imgOut / Ld_Max, 0.0, 1.0);
```

Listing 3.9. MATLAB code: Ferwerda et al. TMO [130] with some of the Durand and Dorsey [117].

Listing 3.9 provides the MATLAB code of the Ferwerda et al. operator [130] with some of the Durand and Dorsey [117] changes. The full code may be found in the file FerwerdaTMO.m. The method takes the following parameters of the display as input: the maximum, Ld_Max, and the adaptation, L_da, display luminance values. The first step is to calculate the luminance adaptation of the input image. Then, the scaling factors for the scotopic and photopic vision are, respectively, calculated using the function TsFerwerda.m and TpFerwerda.m. These functions can be found in the Tmo/util folder.

The k, which simulates the transition from mesopic to photopic range, is subsequently computed and clamped between the range 0 and 1. Finally, Equation (3.11) is applied. The chromaticity for the three RGB colors channels are stored in vec. A normalization step of the output values of the TMO is required to have the final output values between 0 and 1. This is computed by dividing the output values with the maximum luminance of the display device, Ld_Max.

3.2.5 Histogram Adjustment

The classic imaging technique of histogram equalization [144] was extended to handle dynamic range by Larson et al. [215], who also included simulations of aspects of the HVS such as glare, loss of acuity, and color sensitivity.

As the first step, the operator calculates the histogram, H, of the input image in the natural logarithm (ln) space, using n bins. Larson et al. stated that 100 bins are adequate for accurate results. Then, the cumulative histogram, P, is computed as

$$P(x) = \sum_{i=1}^{x} \frac{H(i)}{T}, \qquad T = \sum_{i=1}^{n} H(i), \qquad (3.12)$$

where x is a bin. Note that the cumulative histogram is an integration, while the histogram is its derivative with an appropriate scale:

$$\frac{dP(x)}{dx} = \frac{H(x)}{T\Delta x}, \qquad \Delta x = \frac{\left[\ln(L_{w,\ max}) - \ln(L_{w,\ min})\right]}{n}. \qquad (3.13)$$

The histogram is subsequently equalized. A classic equalization contrast such as

$$\ln(L_d(\mathbf{x})) = \ln(L_{d,\ min}) + P(\ln L_w(\mathbf{x})) \ln(L_{d,\ max}/L_{d,\ min}) \qquad (3.14)$$

exaggerates contrast in large areas of the images, due to compression of the range in areas with few samples, and expansion in very populated ones; see Figure 3.6. This approach can be expressed as

$$\frac{dL_d(\mathbf{x})}{dL_w(\mathbf{x})} \leq \frac{L_d(\mathbf{x})}{L_w(\mathbf{x})}. \qquad (3.15)$$

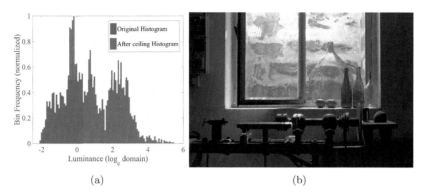

(a)   (b)

Figure 3.6. An example of the histogram adjustment operator by Larson et al. [215] used on the Demijohn HDR image: (a) The histogram of the HDR image. (b) The tone mapped image.

The differentiation of Equation (3.14), using Equation (3.13), and applying Equation (3.15) leads to

$$L_d(\mathbf{x}) \cdot \frac{H(\ln(L_w(\mathbf{x})))}{T\Delta x} \cdot \frac{\ln(L_{d,\,max}/L_{d,\,min})}{L_w(\mathbf{x})} \le \frac{L_d(\mathbf{x})}{L_w(\mathbf{x})},$$

which is reduced to a condition on $H(x)$:

$$H(x) \le c, \qquad \text{where } c = \frac{T\Delta x}{\ln(L_{d,\,max}/L_{d,\,min})}. \tag{3.16}$$

This means that exaggeration may occur when Equation (3.16) is not satisfied. A solution is to truncate $H(x)$, which has to be done iteratively to avoid changes in T and subsequently changes in c. Note that the histogram in Figure 3.6(a) is truncated in $[-2.5, -0.5]$. The operator introduces some mechanism to mimic the HVS, such as limitation of contrast, acuity, and color sensitivity. These are in part inspired by Ferwerda et al.'s [130] work.

In summary, the operator presents a modified histogram equalization for HDR images for range compression that maintains the overall image contrast. Additionally, the method can simulate some aspects of the HVS.

```
%foveal downsanpling
L2 = WardDownsampling(L);

%compute statistics
LMax = max(L2(:));
LMin = min(L2(:));
Llog = log(L2);
L1Max = log(LMax);
L1Min = log(LMin);
```

```
LldMax = log(LdMax);
LldMin = log(LdMin);

%compute the histogram H
H = zeros(nBin, 1);
delta = (LlMax - LlMin) / nBin;

for i=1:nBin
    indx = find(Llog > (delta * (i - 1) + LlMin) & Llog <= (
        delta * i + LlMin)));
    H(i) = numel(indx);
end

%apply histogram ceiling
H = histogram_ceiling(H, delta / (LldMax - LldMin));

%compute P(x)
P = cumsum(H);
P = P / max(P);

%tone map the luminance
L(L > LMax) = LMax;
x = (LlMin:((LlMax - LlMin) / (nBin - 1)):LlMax)';
P_L = interp1(x , P , real(log(L)), 'linear');
Ld  = exp(LldMin + (LldMax - LldMin) * P_L);
Ld  = (Ld - LdMin) / (LdMax - LdMin); %normalize in [0,1]
```

Listing 3.10. MATLAB code: Larson et al. TMO [215].

Listing 3.10 provides the main MATLAB code of the Larson et al. TMO [215]; the full code may be found in the file WardHistAdjTMO.m. The method takes the number of bins, nBin as input. The first step is to scale the HDR input image down to roughly correspond to 1° square pixels using the function WardDownsampling in the folder Tmo/util. This simulates the field of view of an average observer and the eye adaptation [215]. Then, the statistics $L_{w, \, max}$ (LMax) and $L_{w, \, min}$ (LMin) for the downscaled image are computed. At this point, L2, LMin, and LMax are converted to natural logarithmic scale to best capture the luminance population and subjective response over a wide dynamic range. Afterward, the characteristics of the display are defined and converted to natural logarithmic scale. The dynamic range of the display is fixed between 0.01 and 1. The next step is the computation of the histogram H, H, and the cumulative frequency distribution P, P, using Equation (3.12). The histogram is built by finding the elements of L2 that belong to each bin of size delta. After histogram construction, Equation (3.16) is applied to perform the histogram ceiling. The MATLAB code for the histogram ceiling is depicted in Listing 3.11. This follows the pseudo-code from the original paper [215] and can be found in the file histogram_ceiling.m in the Tmo/util folder.

```
function H = histogram_ceiling(H, k)
%tolerance criterion: 2.5% of the ceiling
tolerance = sum(H) * 0.025;
trimmings = 0;
flag = 1;

while((trimmings <= tolerance) & flag)
    trimmings = 0;
    T = sum(H);
    flag = T < tolerance;
    if(T >= tolerance)%compute the new total T
        ceiling = T * k;
        for i=1:length(H)%compute ceiling
            if(H(i) > ceiling)
                trimmings = trimmings + H(i) - ceiling;
                H(i) = ceiling;
            end
        end
    end
end

end
```

Listing 3.11. MATLAB code: Histogram ceiling.

At this point, it is possible to compute the cumulative function P (Equation (3.12)). The variable T maintains the number of samples, and the computation of P corresponds to Equation (3.12). Finally, P is used to tone map the dynamic range of the luminance L.

3.2.6   Time-Dependent Visual Adaptation

While the HVS adapts naturally to large changes in luminance intensity in a scene, this adaptation process is not immediate but takes time proportional to the amount of variation in the luminance levels. Pattanaik et al. [302] proposed a time-dependent TMO to take into account this important effect of the HVS.

The complete pipeline of the model is depicted in Figure 3.7. The first step of the TMO is to convert the RGB input image into luminance values for rods (L_{rod}) and cones (L_{cone}), which are the CIE standard Y' and Y, respectively. Then, the retinal response (R) of rods and cones is calculated as in Hunt's model [178]:

$$R_{\text{rod}}(\mathbf{x}) = B_{\text{rod}} \frac{L_{\text{rod}}(\mathbf{x})^n}{L_{\text{rod}}(\mathbf{x})^n + \sigma_{\text{rod}}^n}, \quad R_{\text{cone}}(\mathbf{x}) = B_{\text{cone}} \frac{L_{\text{cone}}(\mathbf{x})^n}{L_{\text{cone}}(\mathbf{x})^n + \sigma_{\text{cone}}^n},$$
$$(3.17)$$

where $n = 0.73$ [178]. The parameter σ is the half-saturation parameter and is computed for rods and cones as is shown in Equation (3.18) to

Equation (3.19):

$$\sigma_{\text{rod}} = \frac{2.5874 \cdot G_{\text{rod}}}{19000 \cdot j^2 \cdot G_{\text{rod}} + 0.2615 \cdot (1 - j^2)^4 \cdot G_{\text{rod}}^{1/6}}, \qquad (3.18)$$

$$\sigma_{\text{cone}} = \frac{12.9223 \cdot G_{\text{cone}}}{k^4 \cdot G_{\text{cone}} + 0.171 \cdot (1 - j^4)^4 \cdot G_{\text{cone}}^{1/3}},$$

where

$$j = \frac{1}{5 \cdot 10^5 \cdot G_{\text{rod}} + 1}, \text{ and } k = \frac{1}{5 \cdot G_{\text{cone}} + 1}.$$

The value B is the bleaching parameter defined as

$$B_{\text{cone}} = \frac{2 \cdot 10^6}{2 \cdot 10^6 + G_{\text{cone}}}, \qquad B_{\text{rod}} = \frac{0.004}{0.004 + G_{\text{rod}}}. \qquad (3.19)$$

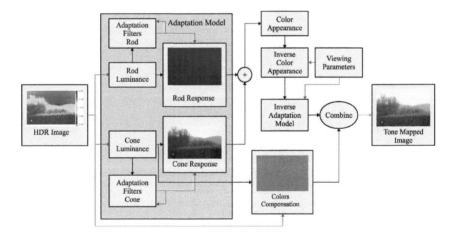

Figure 3.7. The pipeline of the temporal adaptation operator by Pattanaik et al. [302].

G_{cone} and G_{rod} are parameters at adaptation time for a particular luminance value. For a dynamic model, G_{cone} and G_{rod} need to be time dependent: $G_{\text{cone}}(t)$ and $G_{\text{rod}}(t)$. First, the steady state G_{cone} and G_{rod} are computed as one-fifth of the paper-white reflectance patch in the Macbeth checker as suggested by Hunt [178], but other methods are possible such as the one-degree weighting method used in Larson et al. [215]. Note that the authors pointed out that the choice of the initialization method is application dependent. Second, the time dependency is modeled using two exponential filters with output feedback, $1 - e^{\frac{t}{t_0}}$, where $t_{0,\text{rod}} = 150$ ms and

$t_{0,\text{cone}} = 80$ ms. Note that colors are simply modified to take into account range compression in Equation (3.17) as

$$\begin{bmatrix} R'(\mathbf{x}) \\ G'(\mathbf{x}) \\ B'(\mathbf{x}) \end{bmatrix} = \left(\frac{1}{L_{\text{cone}}(\mathbf{x})} \begin{bmatrix} R_{\text{w}}(\mathbf{x}) \\ G_{\text{w}}(\mathbf{x}) \\ B_{\text{w}}(\mathbf{x}) \end{bmatrix} \right)^{S(\mathbf{x})}, \qquad S(\mathbf{x}) = \frac{n B_{\text{cone}} L_{\text{cone}}(\mathbf{x})^n \sigma_{\text{cone}}^n}{(L_{\text{cone}}(\mathbf{x})^n + \sigma_{\text{cone}}^n)^2}.$$

Afterward, a color appearance model (CAM) is applied to compute the scene appearance values Q_{lum} (luminance appearance), Q_{color} (color appearance), Q_{span} (width appearance), and Q_{mid} (midrange appearance) from the RS computed for the rods and cones photoreceptor adaptation model. The final step is to invert the appearance model and the adaptation model to match the viewing condition in ordinary office lighting, where an LDR display is used to visualize the output image.

The operator is a straightforward, global, time-dependent, and fully automatic TMO that lacks some secondary effects of the HVS (glare, acuity, etc.). This model was extended by Ledda et al. [217], who proposed a local model and fully automatic visual adaptation for static images and videos. Additionally, they ran a subjective study for validating the TMO using an HDR monitor [342]. Their results showed a strong correlation between tone mapped images displayed on an LDR monitor and linear HDR images shown on an HDR monitor. A further extension of Pattanaik et al.'s [302] operator was proposed by Irawan et al. [182] who combined the method with the histogram adjustment operator by Larson et al. [215]. Their work simulates visibility in time-varying, HDR scenes for observers with impaired vision. This is achieved using a temporally coherent histogram adjustment method combined with an adaptive TVI function based on the measurements of Naka and Rushton [279].

3.2.7   Adaptive Logarithmic Mapping

Drago et al. [116] presented a global operator based on logarithmic mapping. The authors pointed out how an HDR image tone mapped with logarithm operators using different bases led to different results. Figure 3.8 demonstrates the differences of an image tone mapped with logarithm function using base 2 and using base 10. While the logarithm function with base 10 allows for the maximum compression of high luminance values, the base 2 logarithm function preserves details in areas of dark and medium luminance. Note that neither of these two images provides a satisfying result. Based on these insights, Drago et al. have proposed a TMO that can combine the results of two such images through adaptive adjustment of logarithmic bases depending on the input pixels' radiance.

In particular, they proposed a linear interpolation between $\log_2$ and

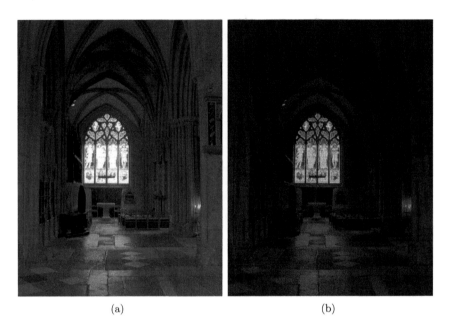

(a)   (b)

Figure 3.8. Results of logarithmic mapping as a TMO using two different bases: (a) Using base 2. (b) Using base 10.

$\log_{10}$ using a smooth interpolation based on Perlin and Hoffer's work [306]:

$$L_{\mathrm{d}}(\mathbf{x}) = \frac{L_{\mathrm{d,\,max}}}{100 \log_{10}(1 + L_{\mathrm{w,\,max}})} \cdot \frac{\log(1 + L_{\mathrm{w}}(\mathbf{x}))}{\log\left(2 + 8\left(\frac{L_{\mathrm{w}}(\mathbf{x})}{L_{\mathrm{w,\,max}}}\right)^{\alpha}\right)}, \qquad (3.20)$$

where $\alpha = \log_{0.5}(b)$, and $b \in [0, 1]$ is a user parameter that adjusts the compression of high values and visibility of details in dark areas. A suggested value is b equal to 0.85; $L_{\mathrm{d,\,max}}$ is the maximum luminance of the display device, where a common value for an LDR display is 100 cd/m^2; see Figure 3.9. The required luminance values are $L_{\mathrm{w}}(\mathbf{x})$ and the maximum luminance of the scene $L_{\mathrm{w,\,max}}$, which need to be scaled by the scene luminance adaptation $L_{\mathrm{w,\,a}}$ and an optional exposure factor. Finally, gamma

```
Lwa = logMean(L);
Lwa = Lwa / ((1.0 + d_b - 0.85)^5);
LMax = max(L(:));

L_s = L / Lwa; %scale by Lwa
LMax_s = LMax / Lwa;

c1 = log(d_b) / log(0.5);
```

```
p1 = (d_Ld_Max / 100.0) / (log10(1 + LMax_s));
p2 = log(1.0 + L_s) ./ log(2.0 + 8.0 * ((L_s / LMax_s).^c1));
Ld = p1 * p2;
```

Listing 3.12. MATLAB code: Drago et al. TMO [116].

correction is applied on the tone mapped data to compensate for the nonlinearity of the display device. This is achieved by deriving a transfer function based on the ITU-R BT.709 standard. The TMO provides a computationally fast mapping that allows a good global range compression. However, its global nature, as with other global methods, does not always allow for the preservation of fine details.

(a)   (b)

Figure 3.9. An example of the adaptive logarithmic mapping, varying p: (a) $p = 0.5$. (b) $p = 0.95$.

Listing 3.12 provides the MATLAB code of Drago et al.'s [116] operator. The full code can be found in the file DragoTMO.m. The method takes as input the maximum luminance of the display device, d_Ld_Max, and the bias parameter, d_b. The first step is to compute the maximum luminance Lmax from the input image and the log luminance, Lwa. Then, L and Lmax are scaled by Lwa. At this point, three variables are computed for the final tone mapping. The first variable c1 corresponds to the parameter α of Equation (3.20). The second, p1, and the third, p2, variables are respectively the left and the right parts of Equation (3.20). These are multiplied together to obtain the tone mapped luminance, Ld.

3.3 Local Operators

Local operators improve the quality of the tone mapped image over global operators by attempting to reproduce both the local and the global contrast. This is achieved by having f, the tone mapping operator, taking into account local statistics of the neighborhood around the pixel being tone mapped. However, neighbors have to be chosen carefully, i.e., choosing only similar pixels or those belonging to the same region. Otherwise, local statistics will be biased producing halos artifacts; see Figure 3.10(b). This typically happens around edges, which separate regions with different intensity and/or color values. Note that halos are sometimes desired when attention needs to be given to a particular area [235], but if the phenomenon is uncontrolled it can produce unpleasant images.

3.3.1 Spatially Nonuniform Scaling

Chiu et al. [90] proposed the first operator that attempted to preserve local contrast. The TMO scales the scene luminance by the average of neighboring pixels, which is defined as

$$L_d(\mathbf{x}) = L_w(\mathbf{x})s(\mathbf{x}), \qquad (3.21)$$

where $s(\mathbf{x})$ is the scaling function that is used to compute the local average of the neighboring pixels, defined as

$$s(\mathbf{x}) = \frac{1}{k(L_w \otimes G_\sigma)(\mathbf{x})}$$

where G_σ is a Gaussian filter and k is a constant that scales the final output. One issue with this operator is that while a small σ value produces a very low contrast image (see Figure 3.10(a)), a high σ value generates halos in the image; see Figure 3.10(b). Halos are caused at edges between very bright and very dark areas, which means that $s(\mathbf{x}) > L_w(\mathbf{x})^{-1}$. To alleviate this, pixel values are clamped to $L_w(\mathbf{x})^{-1}$ if $s(\mathbf{x}) > L_w(\mathbf{x})^{-1}$. At this point s can still have artifacts in the form of steep gradients where $s(\mathbf{x}) = L_w(\mathbf{x})^{-1}$. A solution is to smooth s iteratively with a 3×3 Gaussian filter; see Figure 3.10(d). Finally, the operator masks the remaining halo artifacts by simulating glare, which is modeled by a low-pass filter.

The operator presents the first local solution, but it is quite computationally expensive for alleviating halos (around 1,000 iterations for the

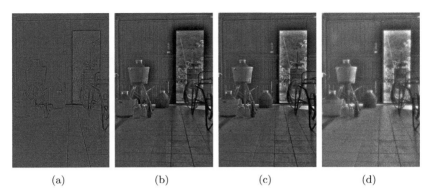

(a)   (b)   (c)   (d)

Figure 3.10. An example of the local TMO introduced by Chiu et al. [90] applied to the Cellar HDR image: (a) The simple operator with $\sigma = 4$. While local contrast is preserved, the global contrast is completely lost, resulting in the flat appearance of the tone mapped image. (b) The simple operator with $\sigma = 32$. In this case, both local and global contrast are kept, but halos are extremely noticeable in the image. (c) The TMO with clamping (2,000 iterations) and $\sigma = 32$. Halos are reduced but not completely removed. (d) The full TMO with glare simulation and $\sigma = 32$ and clamping (2,000 iterations). Note that the glare is used to attempt to mask halos.

smoothing step). There are also many parameters that need to be tuned, and finally halos are reduced but not completely removed by the clamping, smoothing step, and glare.

```
%calculate s
s = RemoveSpecials(1 ./ (c_k * GaussianFilter(L, c_sigma)));

if(c_clamping > 0) %clamp s
    L_inv = RemoveSpecials(1 ./ L);
    indx = find(s >= L_inv);
    s(indx) = L_inv(indx);

    %smoothing s
    H = [0.080 0.113 0.080;...
         0.113 0.227 0.113;...
         0.080 0.113 0.080];

    for i=1:c_clamping
        s = imfilter(s, H, 'replicate');
    end
end

%tone map the luminance
Ld = ChiuGlare(L .* blurred, glare_opt);
```

Listing 3.13. MATLAB code: Chiu et al.'s operator [90].

Listing 3.13 provides the MATLAB code of Chiu et al.'s operator [90]. The full code may be found in the file `ChiuTMO.m`. The method takes as input the parameters of the scaling function, such as the scaling factor `c_k`; `c_sigma` that represents the standard deviation of the Gaussian filter G_σ; `c_clamping` that is the number of iterations for reducing the halo artifacts; and, the parameters (a structure) for the glare filtering `glare_opt`. The first step is to filter the luminance, L, using a Gaussian filter (G_σ in Equation (3.21)) using the function `GaussianFilter.m`, which may be found in the folder `util`. The reciprocal of the filtering result is stored in the variable s. In the case `c_clamping` is higher than 0, the iterative process to reduce halo artifacts is applied to s. This is performed only on the pixels that are still not clamped to 1, after the previous filtering process (smoothing constraint). `indx` stores the index of the pixels in the blurred input HDR luminance that are still above 1. In order to respect the smoothing constraint, the pixels with index, `indx` in s are substituted with the values of the variable `L_inv`.

The variable `L_inv` stores the inverted values of the input HDR luminance, which correspond to the output of the first filtering step in the case $s(\mathbf{x}) > L_w(\mathbf{x})^{-1}$. In this way only these pixel values will be filtered iteratively. The smoothing step is finalized by applying the filter H on the updated s variable. Once the scaling function is computed and stored in s, the dynamic range reduction is obtained by applying Equation (3.21). Glare is computed if the `glare` constant factor is higher than 0. This is performed in an empirical way where only the blooming effect is considered using the function `ChiuGlare.m` in the folder `Tmo/util`.

3.3.2   Photographic Tone Reproduction

A local operator based on photographic principles was presented by Reinhard et al. [323]. This method simulates the burning and dodge effect that photographers have applied for more than a century. In particular, the operator is inspired by the Zonal System presented by Adams [4].

The global component of the operator is a function that mainly compresses high luminance values, and it is defined as

$$L_d(\mathbf{x}) = \frac{L_m(\mathbf{x})}{1 + L_m(\mathbf{x})}, \qquad (3.22)$$

where L_m is the original luminance scaled by $aL_{w,H}^{-1}$, and a is the chosen exposure for film development in the photographic analogy, which can be automatically estimated [321]. $L_{w,H}$ is the logarithmic average that is an approximation of the scene key value. The key value indicates subjectively if the scene is light, normal, or dark. This is used in the zone system for predicting how scene luminance will map to a set of print zones [323].

Note that in Equation (3.22), while high values are compressed, others are scaled linearly. However, Equation (3.22) does not allow bright areas to be burnt out, as a photographer could do to enhance contrast during film development. Therefore, Equation (3.22) can be modified as

$$L_d(\mathbf{x}) = \frac{L_m(\mathbf{x})\left(1 + L_{white}^{-2} L_m(\mathbf{x})\right)}{1 + L_m(\mathbf{x})}, \tag{3.23}$$

where L_{white} is the smallest luminance value that is mapped to white and is equal to $L_{m,\,max}$ by default. If $L_{white} < L_{m,\,max}$, values that are greater than L_{white} are clamped (burnt in the photography analogy).

(a)   (b)

Figure 3.11. An example of the photographic tone reproduction operator by Reinhard et al. [323]: (a) The local operator Equation (3.26) with $\Phi = 4$, $\alpha = 0.05$, and $L_{white} = 10^6$ cd/m^2. (b) The local operator Equation (3.26) with $\Phi = 4$, $\alpha = 0.05$, and L_{white} set to the value of the window luminance. Note that this setting burns out the window providing more control to photographers.

A local operator can be defined for Equation (3.22) and Equation (3.23). This is achieved by finding the largest local area without sharp edges, thus avoiding halo artifacts. This area can be detected by comparing different-sized Gaussian filtered L_m images. If the difference is very small or tends to zero, there is no edge; otherwise there is. The comparison is defined as

$$\left| \frac{L_\sigma(\mathbf{x}) - L_{\sigma+1}(\mathbf{x})}{2^\Phi a \sigma^{-2} + L_\sigma(\mathbf{x})} \right| \leq \epsilon, \tag{3.24}$$

where $L_\sigma(\mathbf{x}) = (L_m \otimes G_\sigma)(\mathbf{x})$ is a Gaussian filtered image at scale σ, and ϵ is a small value greater than zero. Note that the filtered images are normalized so as to be independent of absolute values. The term $2^\Phi a\sigma^{-2}$ avoids singularities, and a and Φ are the key value and the sharpening parameter respectively. Once the largest σ (σ_{max}) that satisfies Equation (3.24) is calculated for each pixel, the global operators can be modified to be local. For example, Equation (3.22) is modified as

$$L_d(\mathbf{x}) = \frac{L_m(\mathbf{x})}{1 + L_{\sigma_{max}}(\mathbf{x})}, \tag{3.25}$$

and similarly for Equation (3.23),

$$L_d(\mathbf{x}) = \frac{L_m(\mathbf{x})\left(1 + L_{white}^{-2}L_m(\mathbf{x})\right)}{1 + L_{\sigma_{max}}(\mathbf{x})}, \tag{3.26}$$

where $L_{\sigma_{max}}(\mathbf{x})$ is the average luminance computed over the largest neighborhood (σ_{max}) around the image pixel. An example of Equation (3.26) can be seen in Figure 3.11 where the burning parameter L_{white} is varied.

The photographic tone reproduction operator is a local operator that preserves edges, avoiding halo artifacts. Another advantage is that it does not need to have calibrated images as an input.

Listing 3.14, Listing 3.15, and Listing 3.16 provide the MATLAB code of the Reinhard et al. [323] TMO. The full code may be found in the file ReinhardTMO.m. The method takes as input the parameters pAlpha, which is the value of exposure of the image a, the smallest luminance that will be mapped to pure white pWhite corresponding to L_{white}, a Boolean value pLocal to decide which operator to apply (0 - global or 1 - local), and the sharpening parameter phi corresponding to Φ as input.

```
%compute logarithmic mean
Lwa = logMean(L);

%scale luminance using alpha and logarithmic mean
Lscaled = (pAlpha * L) / Lwa;
```

Listing 3.14. MATLAB code: Scaling luminance component of Reinhard et al. TMO [323].

The first part of the code computes the scaling luminance step (Listing 3.14). First, the user-set input parameters are verified. Afterward, the luminance is read from the HDR input image and the logarithmic average is computed and stored in Lwa. Finally, the luminance is scaled and stored in L. The local step is performed in the case where the Boolean pLocal variable is set to 1. The scaled luminance, L, is filtered using the MATLAB function `ReinhardGaussianFilter.m` and the condition in Equation (3.24) is used to identify the scale `sMax` (that represents $\sigma_{\max}$) that contains the largest neighborhood around a pixel. Finally, `L_adapt` stores the value of $L_{\sigma_{\max}}(\mathbf{x})$.

```
%compute adaptation
switch pLocal
    case 'global'
        L_adapt = Lscaled;

    case 'local'
        L_adapt = ReinhardFiltering(Lscaled, pAlpha, pPhi);

    case 'bilateral'
        L_adapt = ReinhardBilateralFiltering(Lscaled, pAlpha,
            pPhi);

    otherwise
        L_adapt = Lscaled;
end
```

Listing 3.15. MATLAB code: Local step of Reinhard et al. TMO [323].

In the final step, the `pWhite` is set to the maximum luminance of the HDR input image scaled by $aL_{\mathrm{w,\,H}}^{-1}$ and the final compression of the dynamic range is performed. This is equivalent to either Equation (3.23) or Equation (3.25) depending on whether the global or local operator is used.

```
%range compression
Ld = (Lscaled .* (1 + Lscaled / pWhite^2)) ./ (1 + L_adapt);
```

Listing 3.16. MATLAB code: Last step of Reinhard et al. TMO [323].

An operator that has a similar mechanism to the Photographic Tone Reproduction operator, in terms of preserving edges and avoiding halos, has been presented by Ashikhmin [27]. However, this operator provides two different range compression functions depending on the user's goal, which can be either preserving local contrast or preserving visual contrast.
The function for preserving the local contrast is defined as

$$L_{\mathrm{d}}(\mathbf{x}) = \frac{L_{\mathrm{w}}(\mathbf{x}) f(L_{\mathrm{w,\,a}}(\mathbf{x}))}{L_{\mathrm{w,\,a}}(\mathbf{x})},$$

where f is the tone mapping function, $L_{\mathrm{w,\,a}}(\mathbf{x})$ is the local luminance adaptation, and $L_{\mathrm{w}}(\mathbf{x})$ is the luminance at the pixel location $\mathbf{x}$.
The function for preserving the visual contrast is defined as

$$L_{\mathrm{d}}(\mathbf{x}) = f(L_{\mathrm{w,\,a}}(\mathbf{x})) + \frac{\mathrm{TVI}(f(L_{\mathrm{w,\,a}}(\mathbf{x})))}{\mathrm{TVI}(L_{\mathrm{w,\,a}}(\mathbf{x}))} (L_{\mathrm{w}}(\mathbf{x}) - L_{\mathrm{w,\,a}}(\mathbf{x})),$$

where TVI is a simplified threshold vs. intensity function:

$$\text{TVI}(x) = \begin{cases} \frac{x}{0.0014} & \text{if } x \le 0.0034, \\ 2.4483 + \log(\frac{x}{0.0034})/0.4027 & \text{if } 0.0034 \le x \le 1.0, \\ 16.5630 + \frac{x-1.0}{0.4027} & \text{if } 1.0 \le x \le 7.2444, \\ 32.0693 + \log(\frac{x}{7.2444})/0.0556 & \text{otherwise,} \end{cases} \quad (3.27)$$

where x is a luminance value in cd/m^2. The tone mapping function f is based on the principle that perceptual scale has to be uniform. Therefore, scene luminance is mapped into display luminance according to their relative position in corresponding perceptual scales [27]:

$$L_{\text{d}}(\mathbf{x}) = f\big(L_{\text{w}}(\mathbf{x})\big) = L_{\text{d, max}} \frac{\text{TVI}(L_{\text{w}}(\mathbf{x})) - \text{TVI}(L_{\text{w, min}})}{\text{TVI}(L_{\text{w, max}}) - \text{TVI}(L_{\text{w, min}})},$$

where $L_{\text{d, max}}$ is the maximum luminance of the display device (usually

(a)   (b)

Figure 3.12. A comparison between the TMO by Reinhard et al. [323] and the one by Ashikhmin [27] applied to the Bottles HDR image: (a) The local operator of Reinhard et al. [323]. (b) The local operator of Ashikhmin [27]. Note that details are similarly preserved in both images; the main difference is in the global tone function.

100 cd/m^2). The estimation of the local adaptation luminance $L_{\text{w, a}}(\mathbf{x})$ is based on the principle that balances the two requirements, such as keeping the local contrast signal with reasonable bound while maintaining enough information about image details [27]. This principle leads to averaging over the largest neighborhood that is sufficiently uniform without generating excessive contrast signals (visualized as artifacts). This is a similar mechanism to the one proposed by Reinhard et al., see Figure 3.12 for a comparison between the two operators.

3.4   Frequency-Based/Gradient Domain Operators

Frequency-based/gradient domain operators have the same goal of preserving edges and local contrast as local operators. In this case, as the name implies, range compression is performed in the frequency/gradient domain instead of the spatial domain. Typically, low frequencies (large features) are tone mapped and high frequencies (details) are lightweight processed or left untouched. The main observation for such methods is that edges and local contrast are preserved if and only if a complete separation between large features and details is achieved.

3.4.1   Edge-Aware Filters for Tone Mapping

One of the first operators of this category was proposed by Tumblin and Turk [374], and it is based on low curvature image simplifier (LCIS) filter. The idea is to solve the problem of preserving local contrast without the loss of fine details and texture by replicating the technique used by an artist when painting a scene. In general, an artist would, when drawing a scene, start with a sketch of large features (i.e., boundaries around most important large features) and gradually refine these large features, adding more details, i.e., adding more shading and boundaries.

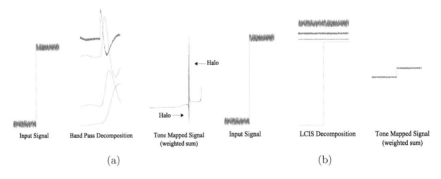

Figure 3.13. Comparison between a band pass and LCIS filters: (a) The band pass filter does not completely separate fine details from large features. (b) LCIS avoids this problem and artifacts are not generated.

A multi-scale application of LCIS filters builds a similar hierarchy by making use of a partial differential equation (PDE) inspired by anisotropic diffusion, and it attempts to compress only the largest features such that the fine details will be preserved.

Figure 3.13(b) shows how an LCIS filters hierarchy works compared to a hierarchy of linear filters. Figure 3.13(a) highlights that linear filters are unable to separate the large features from the fine details. As a consequence of this, some large features may not be compressed because they mixed with the finer details, resulting in halos and characterized by gradient reversal artifacts. Although an LCIS hierarchy is effective, two major drawbacks have limited the adoption of this operator. The first one is that the filtering is computationally expensive because a PDE needs to be solved for each scale. The second drawback is the amount of parameters of the model required in order to achieve the desired look.

(a)   (b)

Figure 3.14. An example of frequencies decomposition on the Bottles HDR image using a bilateral filter: (a) *Base layer*. (b) *Detail layer*.

The bilateral filter. This is a non-linear filter (see Appendix A) that can separate an image into a high frequency image, called *detail layer*, and a low frequency image with preserved edges, called *base layer*. Figure 3.14 shows the decomposition of an HDR image into the two layers.

Durand and Dorsey exploited this property when designing a general and efficient tone mapping framework [118]. This preserves local contrast and is inspired by Tumblin and Turk's work on LCIS [374] but with only a decomposition step, not a hierarchy, in order to have a lightweight model in terms of parameters. The first step of this framework is to decompose an HDR image into luminance and chromaticity. At this point, the luminance in the logarithmic domain is filtered with a bilateral filter obtaining $\log_{10} L_{\text{w, base}}$. The detail layer, $L_{\text{w, det}}$, is calculated by dividing the input HDR luminance by $L_{\text{w, base}}$. This is subsequently tone mapped using a global TMO, $f(x)$, an recombined with the detail layers:

$$L_{\text{d}}(\mathbf{x}) = f\big(L_{\text{w, det}}(\mathbf{x})\big) \cdot L_{\text{w, base}}(\mathbf{x}).$$

The authors employed a simple scaling in the logarithmic domain as TMO, i.e., $\log_{10} f(x) = c \cdot \log_{10} x$ where c is a compression factor. Finally, the tone mapped luminance and the chromaticity are recombined to form the final

(a)   (b)

Figure 3.15. A comparison of tone mapping with and without using the framework proposed by Durand and Dorsey [118] applied to the Bottles HDR image: (a) The tone mapped HDR image using simple scaling in the logarithmic domain. (b) The tone mapped HDR image using the same compression function used in (a) but applied only to the base layer. Note that fine details are enhanced.

tone mapped image. The framework can preserve most of the fine details and can be applied to any global TMO. Figure 3.15 provides an example.

```
%separate detail and base
[Lbase, Ldetail] = BilateralSeparation(L);

log_base = log10(Lbase);
max_log_base = max(log_base(:));
log_detail = log10(Ldetail);
c_factor = log10(target_contrast) / (max_log_base - min(
    log_base(:)));
log_absolute = c_factor * max_log_base;

log_Ld = log_base * c_factor + log_detail   - log_absolute;
Ld = 10.^(log_Ld);
```

Listing 3.17. MATLAB code: The Fast Bilateral Filtering Operator [118].

Listing 3.17 provides the MATLAB code of the bilateral filtering method [118]. The full code may be found in the file DurandTMO.m. The first step of this TMO is to separate the luminance channel, L, into high (Lbase) and low (Ldetail) image frequencies. This is done making use of the MATLAB function BilateralSeparation.m, which can be found in the util folder. Then, the logarithm is applied to Lbase obtaining log_base. This is scaled by a factor, c_factor, that effectively tone maps the base layer. Finally, the details are added back obtaining log_Ld, which is subsequently exponentiated.

Other edge-aware filters. A problem associated with the bilateral filter based decomposition is that the filter kernel can be biased towards one side of a smooth edge in complex spatial intensity transitions in natural images [362]. This leads to over-sharpening during reconstruction that is visible as banding or ringing artifacts. To overcome this limitation, several edge-aware filters have been proposed such as the trilateral filter [92], the weighted least square filter [126], the guided filter [169], L_0 gradient minimization [424], structure preserving smoothing [189], L_1 transform [62], etc.

3.4.2 Gradient Domain Compression

A different approach was proposed by Fattal et al. [129], in which, gradients of the image are modified to achieve range compression. This method is based on the concept that drastic changes in the luminance in an HDR image are due to large magnitude gradients at a certain scale. On the other hand, fine details depend on small magnitude gradients. Therefore, the range can be obtained by attenuating the magnitude of large gradients while keeping (or penalizing less) small gradients to preserve fine details. This method is inspired by Horn's work on recovering reflectance [171].

In the one-dimensional case, this idea is quite straightforward. If an HDR signal $H(x) = \log L_w(x)$ needs to be compressed, derivatives have to be scaled by a spatially varying factor Φ, which modifies the magnitude, only maintaining the direction. This produces a new gradient field $G(x) = H(x)'\Phi(x)$, which is used to reconstruct the compressed signal as

$$\log I(x) = C + \int_0^x G(t)dt,$$

where I is the tone mapped image in the logarithmic domain, and C is an additive constant. Computations are performed in the logarithmic domain because luminance, in this domain, is an approximation of the perceived brightness, and gradients calculated in the logarithmic domain correspond to local contrast ratios in the linear domain. This approach can be extended to the two-dimensional case; e.g., $\mathbf{G}(\mathbf{x}) = \nabla H(\mathbf{x})\Phi(\mathbf{x})$. However, $\mathbf{G}$ is not necessarily integrable because there may be no I such that $\mathbf{G} = \nabla I$. An alternative is to compute a function I whose gradients are closest to $\mathbf{G}$ by minimizing the following:

$$\int\int \|\nabla I(x,y) - \mathbf{G}(x,y)\|^2 dxdy =$$
$$\int\int \left(\frac{\partial I(x,y)}{\partial x} - G_x(x,y)\right)^2 + \left(\frac{\partial I(x,y)}{\partial y} - G_y(x,y)\right)^2 dxdy,$$

which must satisfy the Euler–Lagrange equation, in accordance to the Variational Principle:

$$2\left(\frac{\partial I(x,y)}{\partial x} - G_x(x,y)\right) + 2\left(\frac{\partial I(x,y)}{\partial y} - G_y(x,y)\right) = 0.$$

This equation, once simplified and rearranged, leads to the Poisson's equation:

$$\nabla^2 I - \nabla \cdot \mathbf{G} = 0 \text{ and } L_{\mathrm{d}} = \exp(I). \tag{3.28}$$

The authors suggested to use Neumann boundary conditions, i.e., the derivative in the normal direction to the boundary is zero. In order to solve Equation (3.28), ∇I^2 and $\nabla \cdot \mathbf{G}$ are approximated using finite differences as

$$\nabla I^2 \approx I(x+1,y) + I(x-1,y) + I(x,y+1) - 4I(x,y)$$
$$\nabla \cdot \mathbf{G} \approx G_x(x,y) - G_x(x-1,y) + G_y(x,y) - G_y(x,y-1).$$

The substitution of these approximations in Equation (3.28) yield a large sparse linear system of equation that can be efficiently solved employing a multi-grid method [312].

For Φ, the authors proposed a multi-resolution definition because edges are contained at multiple scales. To avoid halos, the attenuation of edges has to be propagated from the level in which it is detected to the full resolution scale. This is achieved by firstly creating a Gaussian pyramid $(H_0, ...H_d)$, where H_0 is the full resolution HDR image and H_d is at least a 32×32 HDR image. Secondly, gradients are calculated for each level using central differences as

$$\nabla H_i(x,y) = \left(\frac{H_i(x+1,y) - H_k(x-1,y)}{2^{i+1}}, \frac{H_i(x,y+1) - H_k(x,y-1)}{2^{i+1}}\right).$$

At the i-th level, a scaling factor, φ, is defined as

$$\varphi_i(x,y) = \alpha^{1-\beta} \cdot \|\nabla H_k(x,y)\|^{\beta-1},$$

where α is a constant-threshold, assuming that $\beta \leq 1$. While magnitudes above α will be attenuated, the ones below α will be slightly magnified. The final Φ is calculated by propagating the scaling factors φ_i from the coarsest level to the full resolution one:

$$\Phi(x,y) = \Phi_0(x,y), \qquad \Phi_i = \begin{cases} \varphi_d(x,y) & \text{if } i = d, \\ U[\Phi_{i+1}](x,y) \cdot \varphi_i(x,y) & \text{if } i < d, \end{cases}$$

where $U[\cdot]$ is a linear up-sampling operator.

The gradient domain compression operator preserves fine details and avoids halo artifacts by being based on differential analysis; see Figure 3.16. Furthermore, the operator is a very general solution, and it can be employed for the enhancement of LDR images, preserving shadow areas and details.

(a)   (b)

Figure 3.16. An example of tone mapping using the gradient domain operator by
Fattal et al. [129], varying the β parameter applied to Squero HDR image: (a)
$\beta = 0.85$. (b) $\beta = 0.9$. Note that β is directly proportional to global contrast.

3.4.3 Compression and Companding High Dynamic Range Images with Subband Architectures

Li et al. [225] presented a general framework for tone mapping and inverse
tone mapping of HDR images based on multi-scale decomposition. While
the main goal of the algorithm is tone mapping, the framework can also
compress HDR images. A multi-scale decomposition splits a signal $s(x)$
(one-dimensional in this case) into n subbands $b_1(x), \ldots, b_n(x)$ with n filters
$f_1, \ldots, f_n$, in a way that the signal can be reconstructed as

$$s(x) = \sum_{i=1}^{n} b_i(x).$$

Wavelets [356] and Laplacian pyramids [81] are examples of multi-scale
decomposition that can be used in the framework.

The main concept this method uses is based on applying a gain control
to each subband of the image to compress the range. For example, a sigmoid
expands low values and flattens peaks; however, it introduces distortions
that can appear in the final reconstructed signal. In order to avoid such
distortions, a smooth gain map inspired by neurons was proposed. The
first step is to build an activity map, reflecting the fact that the gain of
a neuron is controlled by the level of its neighbors. The activity map is
defined as

$$A_i(\mathbf{x}) = G(\sigma_i) \otimes |B_i(\mathbf{x})|,$$

where $G(\sigma_i)$ is a Gaussian kernel with $\sigma_i = 2^i \sigma_1$, which is proportional to
i, the subband's scale. The activity map is used to calculate the gain map,

which turns gain down where activity is high and vice-versa

$$G_i(\mathbf{x}) = p(A_i\mathbf{x}) = \left(\frac{A_i\mathbf{x} + \epsilon}{\delta_i}\right)^{\gamma-1},$$

where $\gamma \in [0, 1]$ is a compression factor, and ϵ is the noise level that prevents the noise being seen. The equation $\delta_i = \alpha_i \sum_{\mathbf{x}} A_i(\mathbf{x})/(M)$ is the gain control stability level, where M is the number of pixels in the image, and $\alpha_i \in [0.1, 1]$ is a constant related to spatial frequency. Once the gain maps are calculated, subbands can be modified as

$$B_i'(\mathbf{x}) = G_i(\mathbf{x})B_i(\mathbf{x}). \tag{3.29}$$

Note that it is possible to calculate a single activity map for all subbands by pooling all activity maps as

$$A_{ag}(\mathbf{x}) = \sum_{i=1}^{n} A_i(\mathbf{x}).$$

From A_{ag}, a single gain map $G_{ag} = p(A_{ag})$ is calculated for modifying all subbands. The tone mapped image is finally obtained summing all modified subbands B_i'. The compression is applied only to the V channel of an image in the HSV color space [144]. Finally, to avoid over-saturated images, S can be reduced by $\alpha \in [0.5, 1]$. The authors presented comparisons with the fast bilateral filter operator [118], the photographic operator [323], and the gradient domain operator [129]; see Figure 3.17.

The framework can be additionally used for compression, applying expansion after tone mapping. This operation is typically called *companding*. The expansion operation is obtained using a straightforward modification of Equation (3.29) as

$$B_i'(\mathbf{x}) = \frac{B_i(\mathbf{x})}{G_i(\mathbf{x})}.$$

The authors proposed compressing the tone mapped image into JPEG. In this case, a high bit rate is needed (1.5 bpp–4 bpp) with chrominance subsampling disabled to avoid the amplification of JPEG artifacts during expansion, since a simple up-sampling strategy is adopted. This method can be also extended to HDR videos [345] using 3D wavelets and a temporal edge avoiding scheme for avoiding ghosting.

Recently, Fattal [128] and Artusi et al. [26] have presented edge avoiding techniques based on a second generation wavelet (lifting). These approaches define robust prediction operators that weight pixels based on their similarity to the predicted pixel. Specifically, Fattal [128] used an edge-stopping function [229] to define the prediction weights. On the other

(a)   (b)

(c)   (d)

Figure 3.17. A comparison of tone mapping results for the Kitchen HDR image:
(a) The subband architecture using Haar wavelets [225]. (b) The gradient domain
operator [129]. (c) The fast bilateral filter operator [117]. (d) The photographic
operator [323].

hand, Artusi et al. [26] introduced a robust predictor operator, based on
high order reconstruction (HOR) method [163] that intrinsically encloses
an edge-stop mechanism to avoid the influence of pixels from both sides of
an edge. Finally, Paris et al. [298] extended Laplacian pyramids for handling edge-aware image processing to avoid the artificial staircase effects
or ringing effects around sharp edges [79] of some edge-area techniques;

e.g., the bilateral filter. They showed tone mapping applications where fine details are preserved without visual artifacts.

3.5 Segmentation Operators

Another approach to the tone mapping problem is found in segmentation operators. Strong edges and most of local contrast perception is located along the border of large uniform regions. Segmentation operators divide the image into uniform segments, apply a global operator at each segment, and finally merge them. One additional advantage of such a method is that gamut modifications are minimized because a linear operator for each segment sometimes is, in many cases, sufficient.

3.5.1 Segmentation and Adaptive Assimilation for Detail-Preserving

(a)   (b)

Figure 3.18. An example of Tumblin and Rushmeier's TMO [372] using adaptive luminance calculated by the segmentation method of Yee and Pattanaik [428] and applied to the Bottles HDR image: (a) The adaptation luminance calculated using 64 layers. (b) The tone mapped image using the adaptation luminance in (a).

The first segmentation-based TMO was introduced by Yee and Pattanaik [428]. Their operator divides the HDR image into regions and calculates an adaptation luminance for each region. This adaptation luminance can be used as input to a global operator.

The first step of the segmentation is to divide the image into regions, called categories, using a histogram in the $\log_{10}$ logarithmic domain. Contiguous pixels of the same category are grouped together using a flood-fill approach. Finally, small groups are assimilated into a bigger one, obtaining a layer. Small and big groups are defined by two thresholds, tr_{big} and tr_{small}. The segmentation is performed again for n_{max} layers, where the difference is in the size of the bin of the histogram, which is computed as

$$\text{Bin}_{size}(n) = \text{Bin}_{min} + (\text{Bin}_{max} - \text{Bin}_{min})\frac{n}{n_{max} - 1}, \qquad (3.30)$$

where $\text{Bin}_{size}(n)$ is the bin size for the n-th layer, and Bin_{max} and Bin_{min} are, respectively, the maximum and minimum bin size. Once all layers are computed, the adaptation luminance is calculated as

$$L_a(\mathbf{x}) = \exp\left(\frac{1}{n_{max}}\sum_{i=1}^{n_{max}} C_i(\mathbf{x})\right), \qquad (3.31)$$

where $C_i(\mathbf{x})$ is the average log luminance of the group to which a pixel at coordinate $\mathbf{x}$ belongs. The application of L_a to any global TMO helps to preserve edges and thus avoid halos; see Figure 3.18. Note that to avoid banding artifacts a high number of layers (more than 16) is needed.

3.5.2 Lightness Perception in Tone Reproduction

Krawczyk et al. [204] proposed an operator based on the anchoring theory of lightness perception by Gilchrist et al. [142]. This theory states that the highest luminance value, or anchor, in the visual field is perceived as white by the HVS. The perception is affected by relative area. When the highest luminance covers a small area it appears to be self-luminous. To apply lightness theory to complex images, Gilchrist et al. [142] proposed to decompose the image into areas, called frameworks, where the anchoring can be applied.

The first step of the operator is to determine the frameworks. The image histogram is then calculated in the $\log_{10}$ domain. The k-means clustering algorithm is used to determine the centroid, C_i, in the histogram, merging close centroids by a weight averaging based on pixel count. To avoid seams or discontinuity, frameworks are generated with a soft segmentation approach. For each framework, the probability of belonging to it is defined as

$$P_i(\mathbf{x}) = \exp\left(-\frac{(C_i - \log_{10} L_w(\mathbf{x}))^2}{2\sigma^2}\right), \qquad (3.32)$$

where σ is equal to the maximum distance between two frameworks. $P_i(\mathbf{x})$ is smoothed using the bilateral filter to remove small local variations; see Figure 3.19(b) and Figure 3.19(d). Finally, the tone mapped image is calculated as

$$L_d(\mathbf{x}) = \log_{10}(L_w(\mathbf{x})) - \sum_{i=1}^{n} \omega_i P_i(\mathbf{x}), \qquad (3.33)$$

Figure 3.19. An example of the TMO by Krawczyk et al. [204]: (a) and (c) are the frameworks where anchoring is applied. (b) and (d) are, respectively, the smoothed probability maps for (a) and (c). (e) The final tone mapped image is obtained by merging frameworks. (The original HDR image is courtesy of Ahmet Oğuz Akyüz).

where ω_i is the local anchor for the i-th framework, and it is computed as the 95-th percentile of the luminance channel in its framework.

An example of the final tone mapped image can be seen in Figure 3.19. The operator was validated by comparing it with the photographic tone reproduction operator [323] and the fast bilateral filtering operator [118] to test the Gelb effect [142], an illusion related to lightness constancy failure. The result of the experiment showed that the lightness-based operator can reproduce this effect.

Listing 3.18 and Listing 3.19 provide the MATLAB code of the Krawczyk
et al. [204] TMO. The full code may be found in the file KrawczykTMO.m.

```
%compute the image histrogram in log space
[histo, bound, ~] = HistogramHDR(img, 256, 'log10', [], 0, 0, 1
    e-6);

LLog10 = log10(L + 1e-6);

%k-means for computing centroids
[C, totPixels] = KrawczykKMeans(bound, histo);

%partition the image into frameworks
[framework, distance] = KrawczykImagePartition(C, LLog10, bound
    , totPixels);
```

Listing 3.18. MATLAB code: The computation of clusters (using k-means) and
image partition into frameworks of Krawczyk et al. [204] TMO.

Listing 3.18 shows the first steps where the histogram of the luminance
channel, histo, is extracted using the function HistogramHDR.m, which
may be found in the folder util. Then, centroids, C, are computed using
the k-means function KrawczykKMeans.m, which may be found in the folder
Tmo/util. At this point, frameworks are determined using the function
KrawczykImagePartition.m, which may be found in the folder Tmo/util.

```
%compute P_i
sigma = KrawczykMaxDistance(C, bound);
[height, width, ~] = size(img);
sigma_sq_2 = 2 * sigma^2;
K = length(C);
A = zeros(K, 1);
P = zeros(height, width, K);
P_norm = zeros(size(L));
sigma_a_sq_2 = 2 * 0.33^2;

half_dim = min([height, width]) / 2;
for i=1:K
    indx = find(framework == i);
    if(~isempty(indx))
        %compute the articulation of the i-th framework
        maxY = max(LLog10(indx));
        minY = min(LLog10(indx));
        A(i) = 1 - exp(-(maxY - minY)^2 / sigma_a_sq_2);
        %compute the probability P_i
        P(:,:,i) = exp(-(C(i) - LLog10).^2 / sigma_sq_2);
        %spatial processing
        P(:,:,i) = bilateralFilter(P(:,:,i), [], 0, 1, half_dim
            , 0.4);
        %normalization
        P_norm = P_norm + P(:,:,i) * A(i);
```

```
      end
end

%compute probability maps
Y = LLog10;
for i=1:K
      indx = find(framework == i);
      if(~isempty(indx))
            %P_i normalization
            P(:,:,i) = RemoveSpecials(P(:,:,i) * A(i) ./ P_norm);
            %anchoring
            W = MaxQuart(LLog10(indx), 0.95);
            Y = Y - W * P(:,:,i);
      end
end

%clamp in the range [-2, 0]
Y = ClampImg(Y, -2, 0);
%remap values in [0,1]
Ld = (10.^(Y + 2)) / 100;
```

Listing 3.19. MATLAB code: Probability maps calculation step of Krawczyk et al. [204] TMO.

Once frameworks are determined, the probability maps, P, are computed by applying Equation (3.32) to each framework as shown in Listing 3.19. These are multiplied by the articulation factor, A, and their anchoring weight W. Then, Equation (3.33) is applied by subtracting the scaled P to Y, where logarithmic luminance is stored. Finally, the range of Y is clamped and shifted in the range $[-2, 0]$, and the tone mapped luminance, Ld, is obtained by exponentiation.

3.5.3 Interactive Local Manipulation of Tonal Values

A user-based system for modifying tonal values in an HDR/LDR image was presented by Lischinski et al. [229]. The system is based on a brush interface inspired by Levin et al. [222] and Agarwala et al. [9], and it can be seen as a soft-segmentation method. The user specifies which regions in the image require an exposure adjustment by means of brushes; see Figure 3.21.

There are four possible brushes available:

- Basic brush. Sets constraints to pixels covered by the brush, assigning a weight of $w = 1$.

- Luminance brush. Applies a constraint to pixels whose luminance is similar to those covered by the brush. If μ is the mean luminance for pixels under the brush and σ is a brush parameter, a pixel with luminance L is assigned a weight of $w(L) = \exp\left(-(L - \mu)^2 \sigma^{-2}\right)$.

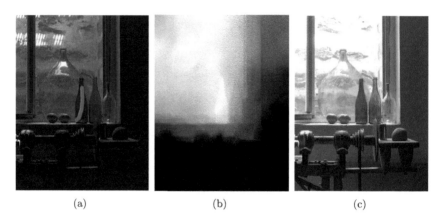

(a)   (b)   (c)

Figure 3.20. An example of the user-based system by Lischinski et al. [229] applied to the Demijohn HDR image: (a) The user decided to reduce the exposure in the under-exposed bottle, so a stroke is drawn in that area. (b) At this point the linear system solver generates a smooth exposure field, keeping strong edges. (c) The final result of the adjustment of the stroke in (a).

- Luma-chrome brush. A brush similar to the Luminance one, but it takes into account luminance and chromaticity.

- Over-exposure brush. Selects all over-exposed pixels that are surrounded by the drawn stroke.

Once the stroke is drawn using the various brushes, the system tries to find a spatially varying exposure function f. This has to satisfy the constraints specified by the user and to take edges into account. In terms of minimization, f can be defined as

$$f = \mathrm{argmin}_f \left(\sum_{\mathbf{x}} w(\mathbf{x}) \big(f(\mathbf{x}) - g(\mathbf{x}) \big)^2 + \lambda \sum_{\mathbf{x}} h(\nabla f, \nabla L') \right),$$

where $L' = \log(L)$ and L is the luminance of an LDR or HDR image. The function $w(\mathbf{x}) \in [0, 1]$ defines constrained pixels, and $g(\mathbf{x})$ is a function that provides the target exposure. The minimization presents a smoothing term that takes into account large gradients and is defined as

$$h(\nabla f, \nabla L') = \frac{|f_x|^2}{|L'_x|^\alpha + \epsilon} + \frac{|f_y|^2}{|L'_y|^\alpha + \epsilon}. \tag{3.34}$$

The variable α defines the sensitivity of the term to the derivatives of the log-luminance image, and ϵ is a small non-zero value used to avoid singularities. The minimum f is calculated by solving a linear system [312]. Following this, the image is adjusted by applying f.

(a)   (b)   (c)

Figure 3.21. An example of the automatic operator by Lischinski et al. [229] applied to the Canal HDR image: (a) The exposure map for each zone. (b) The smoothed exposure map; note that sharp edges are kept. (c) The final tone mapped image.

Finally, the system provides an automatic TMO for preview, which follows the zone system by Adams [4]. The image is segmented in n zones, using Equation (3.35), and the correct exposure is calculated for each zone as

$$n = \left\lceil \log_2(L_{\max}) - \log_2(L_{\min} + 10^{-6}) \right\rceil. \qquad (3.35)$$

These exposures and segments are used as an input for the linear system solver, generating the tone mapped image. An example is shown in Figure 3.21.

The system presents a user-friendly GUI and a complete tool for photographers, artists, and users to intuitively tone map HDR images or enhance LDR images. Furthermore, it can be applied to other tasks such as modifications of the depth of field and spatially varying white balancing.

Listing 3.20 provides the MATLAB code of the Lischinski et al. [229] TMO. The full code may be found in the file LischinskiTMO.m. The method takes as inputs pAlpha, the starting exposure, and pWhite, the maximum output luminance. These parameters are the same as Reinhard et al.'s global operator [323].

```
%number of zones in img
epsilon = 1e-6;
minLLog = log2(min(L(:)) + epsilon);
maxLLog = log2(max(L(:)));
Z = ceil(maxLLog - minLLog);
```

```
%choose the representative Rz for each zone
fstopMap = zeros(size(L));
Lav = logMean(L);
for i=1:Z
    lower_i = 2^(i - 1 + minLLog);
    upper_i = 2^(i + minLLog);
    indx = find(L >= lower_i & L < upper_i);
    if(~isempty(indx)) %apply global ReinhardTMO
        Rz = MaxQuart(L(indx), 0.5);
        Rz_s = (pAlpha * Rz) / Lav; %scale Rz
        f = (Rz_s * (1 + Rz_s / (pWhite^2))) / (Rz_s + 1);
        fstopMap(indx) = log2(f / Rz);
    end
end

%edge-aware filtering
fstopMap = 2.^LischinskiMinimization(log2(L + epsilon),
    fstopMap, 0.07 * ones(size(L)));

imgOut = zeros(size(img));
for i=1:size(img,3)
    imgOut(:,:,i) = img(:,:,i) .* fstopMap;
end
```

Listing 3.20. MATLAB code: The automatic TMO proposed by Lischinski et al. [229].

The algorithm starts by extracting the minimum and maximum luminance values of the input HDR image. The number of zones of the image is computed using Equation (3.35) and stored in Z. The next step is to find a spatial varying exposure function f that is represented by the variable fstopMap. A representative luminance value R_z (R_z) is computed for each zone as median luminance of the pixels in the zone. The global operator of Reinhard et al. [323] is used to map R_z to a target value $f(R_z)$ and is stored in f. Finally, the target exposure is stored in fstopMap. Equation (3.34) is afterward minimized using the Linschinski Minimization.m function that may be found in the Tmo/util folder, and it is a typical solution of a sparse linear system $\mathbf{A} \cdot \mathbf{b} = \mathbf{x}$.

3.5.4   Exposure Fusion

HDR images are usually assembled from a series of LDR images which eventually may be tone mapped. A novel approach that can avoid the tone mapping step was proposed by Mertens et al. [260,261]. This is inspired by the fusion work by Goshtasby [146]. The central concept of this operator is to merge well-exposed pixels from each exposure.

The first step is to analyze each LDR image to determine which pixels can be used during merging. This is achieved by calculating three metrics for each pixel of the i-th image, I_i: contrast, saturation, and well-exposedness. Contrast, $C_i = |\nabla^2 L_i|$, captures the local luminance contrast of the image using the Laplacian operator. Saturation, S_i, is defined as the standard deviation of the red, green, and blue channels:

$$S_i(\mathbf{x}) = \sqrt{\frac{\sum_{j \in \{R,G,B\}} \left(I_{i,j}(\mathbf{x}) - \mu_i(\mathbf{x})\right)^2}{3}}, \text{ and } \mu_i(\mathbf{x}) = \sum_{j \in \{R,G,B\}} \frac{I_{i,j}(\mathbf{x})}{3}.$$

Well-exposedness, E_i, of luminance L_i determines if a pixel is well-exposed in a fuzzy way as

$$E_i(\mathbf{x}) = \exp\left(-\frac{\left(L_i(\mathbf{x}) - \mu_e\right)^2}{2\sigma_e^2}\right), \tag{3.36}$$

where $\mu_e = 0.5$ to assign higher weight to well-exposed pixels, and $\sigma_e = 0.2$ to pick only pixels close to the well-exposed ones. These three metrics are combined together, obtaining a weight $W_i(\mathbf{x})$ that determines the importance of the pixel for that exposure:

$$W_i(\mathbf{x}) = C_i(\mathbf{x})^{\omega_C} \times S_i(\mathbf{x})^{\omega_S} \times E_i(\mathbf{x})^{\omega_E},$$

where i refers to the i-th image, and ω_C, ω_S, and ω_E are exponents that increase the influence of a metric over the others. The N weight maps are normalized such that their sum is equal to one at each pixel position in order to obtain consistent results.

After analysis, the exposures are combined in the final image. To avoid seams and discontinuities, the images are blended using Laplacian pyramids [81]. While the weights are decomposed in Gaussian pyramids as denoted with the operator $\mathbf{G}$, exposure images I_k are decomposed into Laplacian pyramids as denoted with the operator $\mathbf{L}$. Thus, the blending is calculated as

$$\mathbf{L}^l\{I_d\}(\mathbf{x}) = \sum_{i=1}^{n} \mathbf{L}^l\{I_i\}(\mathbf{x}) \times \mathbf{G}^l\{W_i\}(\mathbf{x}) \quad \forall l,$$

where l is the l-th level of the Laplacian/Gaussian pyramid. Finally, the $\mathbf{L}\{I_d\}$ is collapsed, obtaining the final tone mapped image I_d; see Figure 3.22.

The main advantage of this operator is that a user does not need to generate HDR images and to recover the CRF. It also minimizes color shifts that can occur in traditional TMOs because well-exposed pixels are blended without applying a compression function; just a linear scale. These

Figure 3.22. An example of the fusion operator by Mertens et al. [261] applied to the Tree HDR image: (a) The first exposure of the HDR image. (b) The weight map for (a); note that pixels from the tree and the ground have high weights because they are well exposed. (c) The second exposure of the HDR image. (d) The weight map for (c); note that pixels from the sky have high weights because they are well exposed. (e) The fused/tone mapped image using Laplacian pyramids.

two main advantages have made this technique very popular in the software of compact cameras and mobile devices. The success of this operator has sparked many variations of the method such as blending in the bilateral domain [318], entropy-based solutions [78], etc. Furthermore, deghosting and alignment algorithms that work directly while performing the exposure fusion have been proposed [434].

Listing 3.21, Listing 3.22, and Listing 3.23 provide the MATLAB code of the Mertens et al. [261] TMO. The full code may be found in the file MertensTMO.m.

```
if(~isempty(img))
    %convert the HDR image into a imageStack
    [imageStack, ~] = GenerateExposureBracketing(img, 1);
else
    if(isa(imageStack, 'single'))
        imageStack = doubel(imageStack);
    end

    if(isa(imageStack, 'uint8'))
        imageStack = single(imageStack) / 255.0;
    end

    if(isa(imageStack, 'uint16'))
        imageStack = single(imageStack) / 655535.0;
    end
end

%number of images in the stack
[r, c, col, n] = size(stack);
```

Listing 3.21. MATLAB code: Bracketing step of Mertens et al. TMO [261].

The MATLAB implementation (see Listing 3.21) allows for having an HDR image as input (`img`), which is converted into a series of LDR images with different exposures using the function `CreateLDRStackFromHDR.m` in the `IO_stack` folder. A native stack of LDR images, `imageStack`, previously loaded using the function `ReadLDRStack.m`, can be given as input. In this case, `img` has to be set empty.

```
total  = zeros(r, c); %compute weights for each image
weight = ones(r, c, n);
for i=1:n
    if(wE > 0.0)
        weightE = MertensWellExposedness(imageStack(:,:,:,i));
        weight(:,:,i) = weight(:,:,i) .* weightE.^wE;
    end

    if(wC > 0.0)
        L = mean(imageStack(:,:,:,i),3);
        weightC = MertensContrast(L);
        weight(:,:,i) = weight(:,:,i) .* (weightC.^wC);
    end

    if(wS > 0.0)
        weightS = MertensSaturation(imageStack(:,:,:,i));
        weight(:,:,i) = weight(:,:,i) .* (weightS.^wS);
    end

    weight(:,:,i) = weight(:,:,i) + 1e-12;
    total = total + weight(:,:,i);
end
for i=1:n %weights normalization
```

```
    weight(:,:,i) = weight(:,:,i) ./ total;
end
```

Listing 3.22. MATLAB code: Weighting step of Mertens et al. TMO [261].

```
pyrAcc = [];%empty pyramid
for i=1:n
    %Laplacian pyramid: image
    pyrImg = pyrImg3(imageStack(:,:,:,i), @pyrLapGen);
    %Gaussian pyramid: weight
    pyrW   = pyrGaussGen(weight(:,:,i));
    %image times weight (pyramid domain)
    pyrImgW = pyrLstS2OP(pyrImg, pyrW, @pyrMul);
    if(i == 1)
        pyrAcc = pyrImgW;
    else %accumulation
        pyrAcc = pyrLst2OP(tf, pyrImgW, @pyrAdd);
    end
end
imgOut = zeros(r, c, col);
for i=1:col%pyramid reconstruction
    imgOut(:,:,i) = pyrVal(pyrAcc(i));
end
```

Listing 3.23. MATLAB code: Pyramid step of Mertens et al. TMO [261].

Then, each metric is computed for each element of the stack, see Listing 3.22. First, the luminance is extracted from the stack and stored in the variable L. The three metrics for well-exposedness, saturation, and contrast are, respectively, computed using the MATLAB functions MertensWellExposedness.m (Equation (3.36)), MertensSaturation.m (S), and MertensContrast.m (C). They may be found in the Tmo/util folder. Once all metrics are computed, they are used to obtain the final weight maps, weight, that will be used by the blending step. Before this step, the normalization of the weight maps is needed. The total variable is used to store the sum of the n weight maps (n is the number of LDR images) to be used in the last loop of Listing 3.22 for the normalization step. The next step (see Listing 3.23) is to respectively decompose each LDR image and each weight map into a Laplacian decomposition (pyrImg) and a Gaussian one (pyrW) using the MATLAB functions pyrLapGen.m and pyrGauss Gen.m, respectively. The blending is achieved by multiplying pyrImg by pyrW (for one of the n images) and storing the result in pyrImgW. pyrAcc accumulates pyrImgW for each image in the stack. Finally, the pyramid is collapsed into a single image using the MATLAB function pyrVal.m. All MATLAB functions for processing pyramids may be found in the LaplacianPyramid folder.

3.6 Summary

In the last 20 years, several approaches have been proposed to solve the tone mapping problem, as shown earlier in Table 3.1. They have tried to take into account different aspects. These include local contrast reproduction, fine details preservation without introducing halo artifacts, simulation of the HVS behavior, etc.

In spite of a large number of TMOs that have been developed, the tone mapping problem is still an open issue. For example, the reduction of dynamic range can lead to degradation of contrast, which can be solved by applying post-processing techniques [205, 348] or enhancements for binocular vision [427]. The introduction of HDR displays does not fully solve this problem, as there are still images that can exceed the range of current HDR displays, e.g., the picture of a beach during summer can easily exceed 10,000 cd/m^2 on average.

Despite a large number of techniques, dynamic range compression has been mainly tackled on the luminance values of the input HDR image, without properly taking into account how this was affecting color information. Only recently have researchers addressed this problem by proposing the application of color appearance models to the HDR imaging field, a more in-depth understanding of the relationship between contrast and saturation for minimizing the color distortion, etc.

This chapter has given a critical overview of the available techniques to solve the tone mapping problem for still images. Chapter 4 and Chapter 5 respectively present the new trends in tone mapping and how to tone map HDR videos.

4

New Trends in Tone Mapping

The previous chapter presented an overview of tone mapping. The operators presented are typically tailored for a specific goal when reducing the dynamic range of an HDR image such as preserving local/global contrast, matching the perception of the HVS, etc. A more recent trend is to develop generic TMOs that are flexible and can simulate the existing ones [247], or to mix different TMOs [38, 429], or TMOs that adapt the output to be suitable for any display device [242]. In addition, there is now increased attention by researchers in the management of colors. The compression of the HDR input for the low dynamic range of the display has typically been performed only on the luminance range. The color ratio is kept, to be, subsequently, adapted by the compressed luminance range. In the last few years, research has focused on preserving color information. Such techniques take into account how color information can be compressed while maintaining the color appearance of the input image [18, 194, 207, 322] and how the color can be corrected to be mapped into the lower dynamic range of a display device in doing so, reduce the color distortion with respect to the original colors of the HDR input image [241, 311]. At the same time the colors need to be kept inside its gamut range [396].

4.1   A General Approach to Tone Mapping

This section introduces the recent approaches to tone mapping that have been focused on providing either a general and flexible solution or reusing existing methods when reducing the dynamic range of an input HDR image.

4.1.1 Modeling a Generic Tone Mapping Operator

As shown in the previous chapter, several TMOs have been proposed with different features. Nevertheless, evaluation studies (see Chapter 10) have not drawn a conclusion on what constitutes a good TMO.

Mantiuk and Seidel [247] introduced a novel approach to tone mapping. They modeled the processing that is performed by a TMO and proposed a generic three-step TMO. These steps are a per-pixel tone curve, a modulation transfer function, and a color correction step; see Figure 4.1. They demonstrated that this solution may satisfactorily approximate the output of many global and local TMOs.

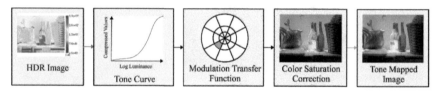

Figure 4.1. Data flow of the generic TMO proposed by Mantiuk and Seidel [247].

The mathematical formula that extrapolates the data flow of the generic TMO for an input HDR image, I_w, is

$$I_{d,i} = MTF(TC(L_w)) \cdot C_i^s \quad \text{where} \quad C_i = \frac{I_{w,i}}{L_w} \quad \forall i \in \{R, G, B\}.$$

The function MTF is the modulation transfer function, TC is a global tone curve applied to each pixel separately, s is the color saturation, and C is the color ratio. The variable $I_{w,i}$ is the i-th color component of the input HDR image, I. Note that color correction is achieved by employing a simple desaturation of colors as seen in the previous chapter.

The authors adopted a four-segment sigmoid function, TC, as a tone curve:

$$TC(L_w) = \begin{cases} 0 & \text{if } L' \leq b - d_l, \\ \frac{1}{2}c\frac{L'-b}{1-a_l(L'-b)} + \frac{1}{2} & \text{if } b - d_l < L' \leq b, \\ \frac{1}{2}c\frac{L'-b}{1-a_h(L'-b)} + \frac{1}{2} & \text{if } b < L' \leq b + d_h, \\ 1 & \text{if } L' > b + d_h, \end{cases}$$

where $L' = \log_{10} L_w$, b is the image brightness adjustment parameter, and c is the contrast parameter. a_l and a_h are parameters that are responsible for the contrast compression for shadows and highlights and are derived as

$$a_l = \frac{cd_l - 1}{d_l} \quad a_h = \frac{cd_h - 1}{d_h},$$

(a) (b)

Figure 4.2. An example of the generic TMO by Mantiuk and Seidel [247]: (a) A tone mapped HDR image using the fast bilateral filtering by Durand and Dorsey [118]. (b) The same tone mapped scene using the generic TMO that emulates the fast bilateral filtering TMO. (Images are courtesy of Rafał Mantiuk.)

where d_l and d_h are the lower and higher mid-tone ranges, respectively.

The TC and the color correction step are sufficient to simulate most global TMOs. In addition, the model simulates spatially varying aspect of the local TMOs through the MTF. This is a one-dimensional function that specifies which spatial frequency to amplify or compress. The authors adopted a linear combination of five parameters and five basis functions to simulate their MTF. In order to simulate a TMO, s and the parameters of both TC and MTF need to be estimated using fitting procedures. The authors proposed to estimate TC's parameters using the Levenberg-Marquardt method and MTF's parameters by solving a linear least-squares problem.

In presenting this generic TMO, the authors highlighted how most state-of-the-art TMOs adopt similar image processing techniques and the differences are the strategy used to choose the set of parameters. Figure 4.2 shows how the method is able to simulate the result of the fast bilateral filtering operator by Durand and Dorsey [118].

4.1.2 Display Adaptive Tone Mapping

As discussed, a TMO is used for reproducing HDR images on a display device with limited dynamic range, typically around 200 : 1. As display devices can have different characteristics such as color gamut, dynamic range, etc, an extra step is often required to manually adjust an image on the display device on which it would be visualized, to ensure a high quality results. This task is often tedious and it typically requires expert skills to achieve good quality results. To solve this problem, Mantiuk et al. [242] proposed a TMO that adaptively adjusts content given the characteristics

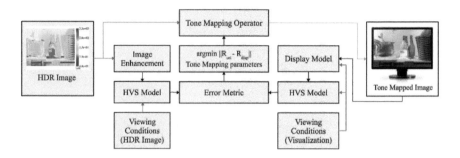

Figure 4.3. The pipeline of the display adaptive TMO proposed by Mantiuk et al. [242].

of a particular display device. Figure 4.3 illustrates the pipeline of the proposed display adaptive TMO.

The authors showed that it is not possible to have a perfect match between the original HDR image and its rendering on the display, when the characteristics of the display are not sufficient to accommodate the dynamic range and color gamut of the original HDR image. However, a trade-off between the preservation of certain image features at the cost of others can be achieved. The choice of the feature to preserve is based on the application, and this process can be automated by designing an appropriate metric. Mantiuk et al. developed this kind of metric through the framework in Figure 4.3. Here a display-adapted image is generated such that it will be a high quality rendering of the original scene. This goal is achieved by comparing the response of the HVS for an image shown on the display R_{disp} and the original HDR input image R_{ori}. In order to compute the responses, an HVS model is integrated with a display model for the image. In the case of the original HDR input image, the HVS model is optionally integrated with an image enhancement model. The parameters of the TMO (i.e., a piece-wise linear function in the logarithmic domain) are computed through the minimization of the difference between R_{ori} and R_{disp}. Some results of the application of the metric presented by Mantiuk et al. [242] for media with different dynamic range are shown in Figure 4.4. Note that the method also works for LDR images if they are scene-referred.

In a similar vein, Wanat and Mantiuk [400] proposed a more general model for retargeting the luminance of an image for matching the appearance under different luminance levels; e.g., retargeting night scenes for very bright displays or vice-versa. This model is based on subjective experiments and it can predict the Purkinje shift (i.e., blue cast at low illumination levels) and color saturation loss.

(a)   (b)   (c)

Figure 4.4. An example of the display adaptive TMO by Mantiuk et al. [242] applied to the Cellar HDR image: (a) A tone mapped image for paper, i.e., 80 cd/m². (b) The same scene in (a) tone mapped for an LDR monitor with 200 cd/m² of peak luminance. (c) The same scene in (a) tone mapped for an HDR monitor with 4,000 cd/m² of peak luminance. Note that images are scaled to be shown on paper. (Images were generated with the help of Rafał Mantiuk.)

4.1.3   Mixing Different Tone Mapping Operators

Another approach to modern tone mapping is to mix the results of different TMOs. The general idea is to make use of the TMOs in certain areas of the image, for which they are known to perform best. An HDR image is thus tone mapped with several TMOs, and all these tone mapped images are then merged together using weights encoding based on how each TMO performs for that area.

There are two main approaches in literature. The first one by Banterle et al. [38, 39] is based on subjective experiments, where the main goal was to determine for each dynamic range zone the TMO that performed the best. From this data, the most suitable TMO for an area of the input HDR image can be determined, i.e., a binary weight map for each operator. The main limitation of this approach is that the authors' results are only for HDR images in the range [0.015, 3000] cd/m², and a similar experiment needs to be run in order to add more TMOs to the set.

The second approach is to use a visual quality metric or a quality index (see Chapter 10) to determine the weights [429]. In this case, the weight map is continuous with values in the range [0, 1] for each TMO. The main limitation of this approach is that a visual quality metric or a quality index is computationally expensive to compute because many filtering operations are required. Therefore, the generation of weights typically requires more computational time than the first approach.

Merging tone mapped images. Similarly to exposure fusion (see Section 3.5.4), these approaches merge the different tone mapped images guided by a weight map for each image. In order to avoid seams, Laplacian pyramids (**L** operator) are used for the tone mapped images and Gaussian pyramids

Figure 4.5. An example of an approach where different TMOs are mixed using TMQI [429]: (a) An HDR image tone mapped using Drago et al.'s operator [116]. (b) The TMQI image, which compares the tone mapped image in (a) against the original HDR image. (c) The same HDR image used as input in (a) tone mapped using Durand and Dorsey's operator [118]. (d) The TMQI image, which compares the tone mapped image in (c) against the original HDR image. (d) The final tone mapped images obtained by mixing using Laplacian pyramids (a) and (b) and their respective TMQI maps used as weights of the mixing.

(**G** operator) are used for weights:

$$\mathbf{L}^l\{I_\mathrm{d}\}(\mathbf{x}) = \sum_{i=1}^n \mathbf{L}^l\{T_{\mathrm{d},i}\}(\mathbf{x}) \times \mathbf{G}^l\{W_i\}(\mathbf{x}) \quad \forall l,$$

where $T_{d,i}$ is the i-th tone mapped image, n is the number of tone mapped images, W_i is the weight of the i-th tone mapped image, and l is the l-th level of the Laplacian/Gaussian pyramid. Finally, the $\mathbf{L}\{I_d\}$ is collapsed, obtaining the final tone mapped image I_d; see Figure 4.5.

4.1.4 Gaze Dependent Tone Mapping

The use of an eye-tracker, which measures the gaze point on a display, combined with an LDR display can be used to provide a tone mapping operator that adapts itself to the area of the image where the user is looking [40, 315]. This approach can simulate how users to perceive real-world environments when eye adaptation occurs. To achieve this, the current gaze point is used by the TMO for computing the adaptation luminance at the gaze point. For example, a simple gaze dependent sigmoid operator could be defined as

$$L_d(\mathbf{x}) = \frac{L_w(\mathbf{x})}{L_w(\mathbf{x}) + f(L_w, \mathbf{y})} \tag{4.1}$$

where $\mathbf{y}$ is the gaze point, and f is a function for computing the adaptation luminance. However, there are a few issues that need to be taken into account in order to produce a high fidelity experience.

Typically, eye-trackers do not measure gaze points with high accuracy. This can lead to the inaccurate computation of the current adaptation luminance and can cause flickering artifacts. To reduce flickering, the adaptation luminance needs to be computed around the area of the current captured gaze position (e.g., 1° of viewing angle) [240].

Another important issue is to ensure the adaptation mechanisms happening in the HVS are appropriately modeled [119,388], such as glare [329], after-images [328], Purkinje color shift [195,400], etc. The model proposed by Jacobs et al. [119], which focused on gaze-dependent visualization, is a quite complete model that simulates global adaptation, mesopic vision, and after-images. Furthermore, the authors ran a series of subjective studies to validate their model and to understand the relationship between the simulated phenomena and the perception of brightness on a LDR display. The results of these experiments showed that mesopic rendering effects (e.g., Purkinjee color shift, and loss of spatial visual acuity) are important cues for brightness perception, but after-images have a low impact.

Although gaze dependent tone mapping has the advantage of providing improved visualization of HDR content on LDR displays, it is has the main limitation of being available to only a single user at a time and an eye tracker is required.

4.2   Color Management in Tone Mapping

Faithful color reproduction is of vital importance for many scientific and industrial applications. Reduction of dynamic range can introduce distortions to the final appearance of colors, i.e., over-saturation. Moreover, TMOs are typically agnostic on the gamut of the visualization display, and the environment where the display is (e.g., dark room, outdoor on a sunny day, etc.).

This section introduces methods for reducing color distortions that happen during tone mapping, and techniques used to manage colors on displays with different color gamuts and environments.

4.2.1   Color Correction

Tone mapping adjusts the contrast relationship in the input HDR image, allowing preservation of details and contrast in the output LDR image. This can often cause a change in color appearance and, to solve this problem, TMOs frequently employ an ad hoc color desaturation step, see Equation (3.2). This does not guarantee that the color appearance is fully preserved, and thus a manual adjustment for each tone mapped image may also be required [241].

Global color correction. Mantiuk et al. [241] performed a series of subjective experiments to quantify and model the correction in color saturation required after tone mapping. Given a tone mapping curve in the luminance domain, the authors wanted to find the chrominance values for the output image that match the input HDR image appearance (with no tone modification). The main aim of Mantiuk et al.'s work was not to compensate for differences in viewing conditions between the display and the real-world scenes, as occurs in the color appearance models, but to preserve the appearance of the reference image without tone modification when shown on the same display. The reference image is chosen as the best exposure of an HDR image that takes into account the environment at which it is displayed and the display model.

The results allow the relationship between the contrast factor c and the saturation factor s to be quantified and approximated by a power function, which is defined as

$$s(c) = \frac{(1+k_1)c^{k_2}}{1+k_1 c^{k_2}} \qquad c(L_{\mathrm{w}}) = \frac{d}{dL'}f(L'),$$

where f is the TMO, and $L' = \log_{10} L_{\mathrm{w}}$. k_1 and k_2 are parameters that were fitted using least squares fitting for two different goals: non-linear color correction and luminance preserving correction. In the case of non-linear color correction, k_1 and k_2 are 1.6774 and 0.9925, respectively. For

Figure 4.6. An example of the color correction technique for tone mapping: (a) The reference HDR image (scaled, gamma corrected, and clamped for visualization purposes). (b) The image in (a) tone mapped with the Reinhard et al.'s global operator [323]. (c) The image in (b) with manual color correction using Equation (4.2) and $s = 0.75$ (found via a trial and error approach). (d) The image in (b) with Mantiuk et al.'s automatic color correction using Equation (4.2). (e) The image in (b) with Pouli et al.'s color correction [311].

luminance preserving correction, k_1 and k_2 are 2.3892 and 0.8552, respectively. The formula can be easily integrated in existing TMOs, where the preservation of the color ratio takes into account the saturation factor s as

$$I_{\mathrm{d},i}(\mathbf{x}) = \left(\frac{I_{\mathrm{w},i}(\mathbf{x})}{L_{\mathrm{w}}(\mathbf{x})}\right)^{s} L_{\mathrm{d}}(\mathbf{x}), \quad \forall i \in \{R, G, B\}. \qquad (4.2)$$

As noted by the authors, the main drawback of Equation (4.2) is that it alters the resulting luminance for $s \neq 1$ and for colors different from gray. The authors suggested a different formula that preserves luminance and involves only linear interpolation between chromatic and achromatic colors:

$$I_{\mathrm{d},i}(\mathbf{x}) = \left(\left(\frac{I_{\mathrm{w},i}(\mathbf{x})}{L_{\mathrm{d}}(\mathbf{x})} - 1\right)s + 1\right)L_{\mathrm{w}}(\mathbf{x}), \quad \forall i \in \{R, G, B\}. \qquad (4.3)$$

However, Equation (4.3) can lead to hue shifts, especially for red and blue colors. Therefore, the choice between Equation (4.2) and Equation (4.3) depends on the final reproduction goal; preserving luminance or hue respectively.

Figure 4.6(d) shows a result of the color correction techniques proposed by Mantiuk et al. [241] compared with the traditional color correction used in the tone mapping problem, making use of a different saturation value S 0.3 (c) and 1.0 (b), respectively.

```
L = lum(img);
imgOut = zeros(size(img));

for i=1:size(img, 3);
    M = (img(:,:,i) ./ L);
    Mc = (M - 1.0) .* cc_s + 1.0;
    imgOut(:,:,i) = Mc .* L;
end

imgOut = RemoveSpecials(imgOut);
```

Listing 4.1. MATLAB code: linear color correction (i.e., either desaturation or saturation) for images (ColorCorrectionLinear.m).

Listing 4.1 applies Equation (4.3) in a straightforward way. The full code can be found in the file ColorCorrectionLinear.m in the folder ColorCorrection. Note that the correction value, cc_s (s), can be a single channel image per pixel correction.

General saturation correction. Although Mantiuk et al.'s method [241] allows parameters prediction for both Equation (4.2) and Equation (4.3), the tone compression function has to be global. Pouli et al. [311] observed that the amount of desaturation can be inferred from the non-linearity applied by the tone curve, irrespective of whether the tone reproduction operator was spatially varying or not. Starting from the original HDR image, I_w, and the color-distorted tone mapped image, I_d, their main goal is to match I_w color appearance in terms of hue and saturation, while preserving luminance values from I_d.

In order to keep hue uniform, the image is converted to XYZ and subsequently to IPT color space for image representation; see Appendix B. To directly manipulate chroma and hue, IPT color values are converted into a cylindrical color space, ICh (I lightness, C chroma, and h hue), which is similar to CIE LCh. The conversion is defined as

$$\begin{bmatrix} I \\ C \\ h \end{bmatrix} = \begin{bmatrix} I \\ \sqrt{P^2 + T^2} \\ \tan^{-1}(P/T) \end{bmatrix}$$

In order to achieve a high quality color correction, the lightness of the tone mapped image needs to be preserved, i.e., the I channel is not further processed. Additionally, the hue distortion needs to be minimized while matching the saturation of the tone mapped image to the one of the original HDR image. Since the hue of the tone mapped image, h_d, may be distorted by gamut clipping during tone-mapping, it is set to the hue of the original HDR image, h_w. Saturation, s, is computed through a formula that mimics the human perception more closely [233]:

$$s(C, I) = \frac{C}{\sqrt{C^2 + I^2}}. \tag{4.4}$$

The tone mapping step compresses luminance in a nonlinear manner without modifying the chromatic information. Since saturation depends on lightness, modifying luminance (or lightness) changing the saturation of the tone mapped image leads to an oversaturated effect [311]. The nonlinear luminance mapping during tone-mapping is the cause of an increment of relative luminance for several pixels of the tone mapped image when compared with their surrounding pixels [311]. To deal with this mismatch, the chroma C_d of the tone mapped image is scaled to approximately what it would be if the original HDR image had been tone mapped in the CIE $L^*C^*h^*$ color space; see Appendix B.

$$C_d' = C_d \frac{I_w}{I_d}. \tag{4.5}$$

Then, based on Equation (4.4), the ratio r between the saturation of the original and tone mapped images is computed:

$$r = \frac{s(C_w, I_w)}{s(C_d', I_d)}. \tag{4.6}$$

This ratio is then applied to chroma C_d' to find the chroma-corrected C_c for the tone mapped image:

$$C_c = rC_d' = r\frac{I_w}{I_d}C_d.$$

The hue of the tone mapped image is reset by copying values from the HDR images ($h_c = h_w$). The chroma-corrected C_c is combined with the lightness channel of the tone mapped image $I_c = I_d$ to produce the final corrected result, which can then be converted back to RGB; see Figure 4.6(e). Subjective experiments showed that this color correction method matches the impression of saturation between tone mapped images and their HDR originals measurably better than other approaches such as the one proposed by Mantiuk et al. [241].

```matlab
%normalize both images
imgHDR = imgHDR / max(imgHDR(:));
imgTMO = imgTMO / max(imgTMO(:));

%conversion from RGB to XYZ
imgTMO_XYZ = ConvertRGBtoXYZ(imgTMO, 0);
imgHDR_XYZ = ConvertRGBtoXYZ(imgHDR, 0);
%conversion from XYZ to IPT
imgTMO_IPT = ConvertXYZtoIPT(imgTMO_XYZ, 0);
imgHDR_IPT = ConvertXYZtoIPT(imgHDR_XYZ, 0);
%conversion from IPT to ICh
imgTMO_ICh = ConvertIPTtoICh(imgTMO_IPT, 0);
imgHDR_ICh = ConvertIPTtoICh(imgHDR_IPT, 0);

%the main algorithm
C_TMO_prime = imgTMO_IPT(:,:,2) .* imgHDR_ICh(:,:,1) ./
    imgTMO_ICh(:,:,1);
r1 = SaturationPouli(imgHDR_ICh(:,:,2), imgHDR_ICh(:,:,1));
r2 = SaturationPouli(C_TMO_prime, imgTMO_ICh(:,:,1));
r = r1 ./ r2;
imgTMO_ICh(:,:,2) = r .* C_TMO_prime; %final scale
imgTMO_ICh(:,:,3) = imgHDR_ICh(:,:,3); %HDR's hue

%conversion from IPT to XYZ
imgTMO_c_IPT = ConvertIPTtoICh(imgTMO_ICh, 1);
%conversion from IPT to XYZ
imgTMO_c_XYZ = ConvertXYZtoIPT(imgTMO_c_IPT, 1);
%conversion from XYZ to RGB
imgTMO_c = ConvertRGBtoXYZ(imgTMO_c_XYZ, 1);
```

Listing 4.2. MATLAB code: Pouli's color correction for tone mapped images (`ColorCorrectionPouli.m`).

Listing 4.2 provides the MATLAB code of Pouli et al.'s color correction. The full code can be found in the file `ColorCorrectionPouli.m` in the folder `ColorCorrection`. The function takes as input the tone mapped image, `imgTMO` and the original HDR image, `imgHDR`, which are supposed to be in a *RGB* color space with BT.709 primaries. The first step of the algorithm is to normalize both images. Subsequently, both images are converted from *RGB* to *ICh* using color conversion functions that can be found in the folder `ColorSpace`. At this point, the chroma of the tone mapped images is scaled as in Equation (4.5) and stored in the variable `C_TMO_prime`. Then, the second scaling factor, Equation (4.6), is computed, multiplied by `C_TMO_prime`, and stored in the output tone mapped image, `imgTMO_ICh(:,:,2)`. The hue of the HDR image, `imgHDR_ICh(:,:,3)` is assigned to the hue of the corrected tone mapped image. Finally, this image is converted from *ICh* color space to *RGB* color space obtaining `img_TMOc`. This is achieved by reversing the order of the function calls at the beginning.

4.2.2 Color Gamut

The color correction techniques described in Section 4.2.1 provide solutions to reduce the color saturation that produces a different color appearance between the original HDR image and the tone mapped image. However, these methods do not take into account the destination gamut where the tone mapped image is finally displayed. As a consequence of this, pixels of the tone mapped image may be left out from the display gamut. Moreover, if pixels are mapped into the display gamut, strong hue shift and luminance distortion may result. In standard imaging, the process to fit the gamut of an image into the gamut of a display is called *gamut mapping*, and there are several methods for achieving this [273].

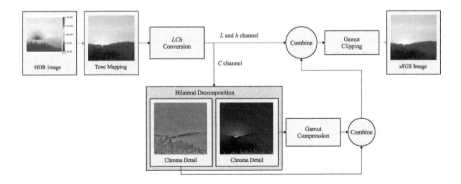

Figure 4.7. The HDR gamut mapping framework proposed by Šikudová et al. [396].

Local HDR color gamut. In the HDR domain, a general solution for this problem was proposed by Šikudová et al. [396], which integrates tone mapping with gamut mapping; see Figure 4.7. This makes it possible to fit colors, after dynamic range compression, within the target display gamut.

The first step is to tone map an input HDR image in the XYZ color space, where range compression is applied to the luminance channel, L_w, only obtaining L_d. Subsequently, L_d is substituted to Y in the original XYZ values of the input image, and it is normalized so the maximum value is 100 cd/m^2. Gamut boundaries are then computed for both the input HDR image and the target display. This step identifies if pixels of the source gamut are either within or out of the destination gamut, and computes the mapping directions needed for the gamut mapping function.

The algorithm employs the CIE LCh color space [346] for minimizing hue and chroma shifts; see Appendix B for converting an XYZ color value into a CIE LCh one. To preserve edges in C, the bilateral filtering is employed to decompose the image into detail and a base layers; see Section 3.4.1. Only the base layer is compressed, and the detail layer is added back.

The C base layer compression can be performed using two approaches. The first one, named the hue-specific method, can maximize the use of the available gamut while fully compressing the chroma. However, this has a high computational complexity. A simpler and quality-efficient chroma compression method, *the global approach*, has been proposed, which has a lower computational complexity. The global approach relies on the observation that the dynamic range of the chroma channel is not extremely high, when compared to the dynamic range of the destination gamut. This suggests that linear compression may be enough. As indicated above, the chroma compression is applied only to the base layer of the chroma channel, $C_{\text{base},h}$, as follows:

$$C'_{\text{base},h} = C_{\text{base},h} \min_{h \in [0°,359°]} \left[\frac{\text{Cusp}_{d,h}}{\text{Cusp}_{s,h}} \right], \tag{4.7}$$

where $\text{Cusp}_{s,h}$ and $\text{Cusp}_{d,h}$ are, respectively, the maximum chroma values for the source and destination gamuts at a specific hue angle h. In other words, the compression is performed at each hue angle h with an incremental step of 1 degree. Note that only the source gamut is compressed and only when $\min\left(\text{Cusp}_{d,h}/\text{Cusp}_{s,h}\right) < 1$. The chroma compression not only avoids hue shift when compared to clipping techniques, but it also reduces the oversaturated appearance typical of the tone-mapping step. Using the full chroma range of the source gamut may produce extremely compressed chroma results. This is typically due to the use of a few outlier pixels with extremely high chroma values. This problem can be avoided by using a percentile approach that eliminates these outlier pixels [396]. After compression, a smooth gamut clipping is applied because an independent processing of lightness and chroma channels cannot guarantee that all pixels are within the destination gamut.

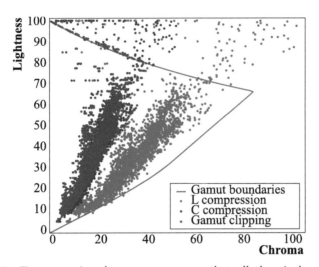

Figure 4.8. Tone mapping does not guarantee that all the pixels of the tone mapped image are within the destination gamut where the image will be visualized. Pixels may be within the destination gamut only for the lightness channel L^*; however, their chroma channel may still be out of gamut. (Image courtesy and copyright of Tania Pouli.)

Figure 4.8 shows this problem. Tone mapping guarantees that the maximum value of lightness is 100 cd/m², but a few pixels may be still out of the destination gamut in their chroma direction. This depends on how C was compressed. Finally, the clipped corrected version of the CIE LCh image is converted into the $sRGB$ color space.

The authors also proposed a lightness compression strategy, integrated within the proposed framework. In this case, the lightness component $L^\star$ is also filtered with a bilateral filter (see Appendix A) and the compression is performed on its base layer, while the details are added back after the compression step. The main advantage over the existing color correction techniques is that the hue shift and luminance distortions introduced by their framework is limited.

4.2.3 Color Appearance Models

A color appearance model (CAM) describes how the HVS is adapting to different environments and how it will perceive the images in these different environments [124]. Therefore, its main focus is on color.

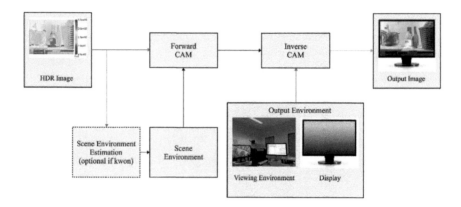

Figure 4.9. An example of a general CAM.

Figure 4.9 shows an example of a general pipeline of a CAM. The first step of a CAM is to take into account the chromatic adaptation, which is the mechanism that enables the HVS to adapt to the dominant colors of illumination. Afterward, the color appearance attributes that are correlated to perceptual attributes of color such as chroma, hue, etc., are computed. Once the appearance attributes are known, the CAM can be reversed and by using the viewing environment conditions as input, the colors in this new environment can be calculated. The differences between viewing environments in which an image is displayed plays an important role on how the color appearance is perceived by the observer. A number of methods have been suggested to take this into account.

Pre-processing HDR images. Akyüz and Reinhard [18] proposed a scheme to integrate the use of a CAM into the context of tone mapping. Figure 4.10 demonstrates this method. Here a CAM is first applied to the original HDR input image to predict the color appearance attributes. Then, the color appearance attributes, together with the new viewing conditions of the environment where the HDR input image will be displayed, are given as input to the reversed CAM that predicts the color stimulus of the HDR input image for the new viewing environment. Before applying the dynamic range compression step via tone mapping, the output of the reversed CAM is reset with the original luminance of the HDR input image while the chromatic information is retained. This final step is necessary to remove any luminance compression that the CAM may have applied, e.g., the CIECAM02 model [272] has a non-linear response compression step, which is not fully inverted in the backward model.

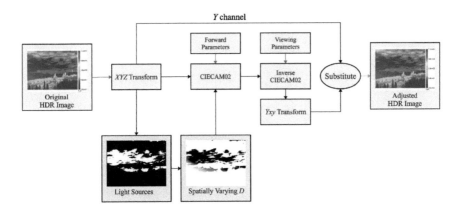

Figure 4.10. Scheme proposed by Akyüz and Reinhard [18] on how to integrate the use of CAM in the context of the tone mapping problem.

Color appearance and tone mapping. Several CAMs have been proposed for the reproduction of HDR images. Typically, they work in a color space for improving the simulation of HVS mechanisms, such as LMS (for cones' responses), IPT (for processing in a uniform color space), etc. Another common feature is to use a sigmoid function, Equation (1.1), for compressing the dynamic range mimicking the process that happens within the HVS.

Pattanaik et al. [301] proposed one of the first spatially varying color appearance models (CAMs) for the reproduction of HDR images. This CAM processes images in LMS color space for cones' responses and rods' responses. It is based on previous psychophysical research arranged in a coherent framework. The key idea is that the visual processing can be described by filtering at different scales [421] using a stack of difference of Gaussian images. This models the CSFs as described in Watson and Solomon [413]. The model, due to its local nature, can simulate many aspects of the HVS, such as visual acuity, change in threshold visibility, color discrimination, and colorfulness. Figure 4.11(a) shows an example of this CAM. However, halos can occur if the starting kernel for the decomposition is not chosen accurately. Similar algorithms to this CAM are multi-scale methods based on Retinex theory [212], such as Raham's work [316], and its extensions to HDR imaging [264].

iCAM06 is a CAM based on the iCAM02 framework [125]. It was proposed by Kuang et al. [207] for tone mapping. The method works in the XYZ color space. It also employs the IPT color space [178] in order to predict the Hunt effect, i.e., an increase in luminance level results in an increase in perceived colorfulness. The method decomposes the image

(a)   (b)   (c)

Figure 4.11. An example of different CAMs applied to the Canal HDR image:
(a) The method by Pattanaik et al. [301]. (b) The method by Kuang et al. [207].
(c) The method by Kim and Kautz et al. [194].

into a details layer (high frequencies) and a base layer (high frequencies)
using the bilateral filter. This allows the processing of the detail layer to
simulate Stevens' effect, which predicts the increase of the local contrast
when luminance levels are increased; see Figure 4.11(b) for an example.

Kim and Kautz [194] conducted a series of subjective experiments to
acquire appearance data for different luminance levels up to $16{,}860 \text{ cd}/\text{m}^2$,
i.e., extended luminance. These experiments were consistent with the literature but they provided data under high-luminance conditions which have
been never tested before. From these data, the authors proposed a new
global color appearance model, which performs well at high luminance levels. The model works in the LMS color space, and it employs a tailored
sigmoid function for range compression; see Figure 4.11(c) for an example.   When calibrated images need to be visualized on a wide range of
monitors and environmental conditions, the CAM by Reinhard et al. [322]
is effective. In fact, most CAMs [194, 207, 272] use a single value for describing the viewing environment or the display. However, a single value
may not be precise enough. Reinhard et al.'s CAM employs more parameters such as white point, maximum luminance, adapting white point,
and adapting luminance for characterizing the scene where the image was
taken, the viewing environment, and the viewing display. Furthermore,
the method can model local aspects of lightness perception by exploiting
the median-cut algorithm [108]; see Section 7.2.1. All these aspects lead
to a precise appearance reproduction removing the guesswork when setting

environmental and display parameters.

4.3 Summary

Despite more than 20 years of research into tone mapping there is still much work to be done. This chapter has described recent developments. These are focusing on three important aspects of TMOs: the need to preserve color better during tone mapping; how TMOs can be simplified to perhaps provide a general solution for which more targeted operators can be built; and, how to best take into account the display and ambient lighting conditions where the tone mapped content will actually be viewed.

It is clear that more work is still needed though. As our understanding of the HVS grows, so new TMOs will be able recreate what the human perceives more accurately. Of particular importance, with the rapid growth of mobile devices, including tablets and smart phones as a preferred platform for watching video content, will be how TMOs can continue to deliver an enhanced viewing experience for a wide range of displays and with dynamically changing ambient lighting conditions as more content is watched "on the go".

5
HDR Video Tone Mapping

A straightforward approach for tone mapping HDR videos is to apply a TMO of choice to each frame of an input HDR video independently. However, this approach typically performs poorly because global and local statistics can abruptly change in a video sequence; e.g., a very bright object appearing in a subsequent frame as shown in Figure 5.1. This means that flickering artifacts occur, making watching the video unpleasant.

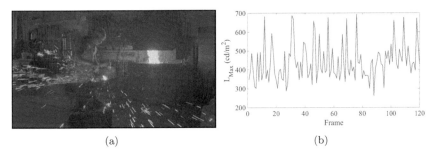

(a)   (b)

Figure 5.1. HDR video statistics for the first 120 frames of the *Smith Hammering* HDR video sequence [137]: (a) A tone mapped frame of the sequence. (b) The plot of the maximum luminance of each frame. (The original HDR video is courtesy of Jan Fröhlich.)

As seen in previous chapters, TMOs typically employ global and local statistics in order to achieve high quality dynamic range compression, i.e., fitting as much relevant data as possible into the available 8 bits. A common solution to tone mapping HDR videos is to anchor global statistics to a frame of choice; e.g., the darkest or the brightest frames, and use the statistics of the chosen frame for all other frames. Although this solution is a rapid fix to global flickering, it does not create high quality results. For example, if the anchor frame is set to be the brightest frame most of frames will have a dark appearance, and vice versa; see Figure 5.2.

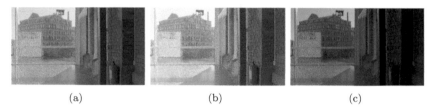

(a)   (b)   (c)

Figure 5.2. HDR video static tone mapping applied to the 22nd frame of the *Hallway2* HDR video sequence [148] (a dark frame): (a) The frame tone mapped with Drago et al.'s operator [116] using the frame statistics. (b) The frame tone mapped in (a) with the statistics of the darkest frame in the sequence; note that the appearance is brighter than in (a). (c) The frame tone mapped in (a) with the statistics of the brightest frame in the sequence; note that the appearance is brighter than in (a). (The original HDR video is courtesy of Jonas Unger.)

Moreover, anchoring to the brightest frame may produce very low contrast tone mapped videos that sub-utilize the available 8 bits.

5.1 Temporal Artifacts

The use of tone mapping in HDR video applications needs to take into account different kind of temporal artifacts that can appear when processing frames of an HDR video independently. Global flickering, as mentioned previously, is just one of the more noticeable artifacts, but a number of others can also occur. Two studies in the field [69, 123] have categorized the main types of artifacts as:

- Flickering Artifacts (FA). This occurs when there are abrupt changes in lighting between frames. This is due to the changes in the dynamic range between frames that results in large discrepancies between the frames' computed luminance statistics such as logarithmic mean, minimum, and maximum. As result the brightness of each tone mapped frame can change quite noticeably from one video frame to the next, leading to flicker in the final video stream [196]. Flickering artifacts may be further classified as global (GFA) and local (LFA) flickering artifacts. The former takes place globally on the whole frame, while the latter only affects small areas of the frame.

- Temporal Brightness Incoherency (TBI). This artifact occurs when not representing illumination changes in HDR videos. It can happen when a TMO may try to find the best exposure for reducing the dynamic range, such as a video sequence in which a light dims; a

TMO that suffers from TBI would not show the dimming but rather a nearly constant illumination.

- Temporal Object Incoherency (TOI). This appears when the brightness of a person/object, which is stable in the HDR sequence, varies in the tone mapped video. For example, a person becomes brighter (or vice-versa) even though the lighting in the scene has not changed. This artifact is due to the adaptation process of a TMO (if it has one), and it is similar to the behavior of cameras that record videos using automatic exposure.

- Temporal Hue Incoherency (THI). This is a variation of the color perception of a person/object in the scene due to color clipping. It is similar to TOI but for color.

- Ghosting (G). This appears, as in the case of assembling an HDR image (see Section 2.1.1), when an area of a frame exhibits partially a person/object that was in that area in the previous frame. This happens when a method, which enforces temporal coherence using, for example, optical flow, cannot estimate the motion vectors correctly due to bad matches; e.g., due to occlusions or appearance/disappearance of on-screen objects.

- Temporal Noise (TN). An HDR video tends to have higher levels of noise due to the current capturing technology; e.g., varying the ISO of the video camera spatially or temporally (see Section 2.1.2). This happens especially in the very dark areas of the captured video. However, many TMOs, due to their design, are not noise-aware so they can boost the brightness of dark areas making noise more noticeably. The noise is also, typically, not stable and it further compounds the issue by creating local flickering in dark areas.

5.2 HDR Videos MATLAB Framework

The HDR video framework consists of functions for reading HDR videos and can be found in the folder IO_video. Currently, this supports HDR videos as a folder containing individual frames stored as HDR images (.hdr, .exr, .pfm, etc.). However, the framework is designed such that it can be easily extended to include native video formats. To read a video, the hdrvread.m function needs to be invoked. This takes as input the name of either a folder, which contains HDR frames stored as images, or the filename of a native HDR video (if supported). This function returns a structure, hdrv, containing information to open/close the video stream,

read/write frames, etc. To access to a video frame, the first step is to open
the video stream by passing an `hdrv` to the function `hdrvopen.m`, which
will return an updated `hdrv` where the video stream is open. Finally, a call
to the function `hdrvGetFrame` with the opened `hdrv` and the index to the
frame to extract, `frameCounter`, will return the decoded frames one at a
time. Once the desired frames are extracted, the open video stream needs
to be closed using the function `hdrvclose.m`.

```
nBins = 4096; %histogram's number of bins
stats_v = zeros(hdrv.totalFrames, 6);
hists_v = zeros(hdrv.totalFrames, nBins);
%open the HDR video stream
hdrv = hdrvopen(hdrv);
for i=1:hdrv.totalFrames
    %fetch the current frame
    [frame, hdrv] = hdrvGetFrame(hdrv, i);
    %compute luminance
    L = lum(RemoveSpecials(frame));
    %avoid negative values
    indx = find(L >= 0.0);
    if(~isempty(indx))
        stats_v(i, 1) = min(L(indx));
        stats_v(i, 2) = max(L(indx));
        stats_v(i, 3) = MaxQuart(L(indx), 1 - percentile);
        stats_v(i, 4) = MaxQuart(L(indx), percentile);
        stats_v(i, 5) = mean(L(indx));
        stats_v(i, 6) = logMean(L(indx));
        hists_v(i, :) = HistogramHDR(L(indx), nBins, 'log10',
            [-6, 6], 0, 0, 1e-6);
    end
end
%close the HDR video stream
hdrv = hdrvclose(hdrv);
```

Listing 5.1. MATLAB code: Extraction of HDR video statistics per frame.

A useful operation for many video TMOs is to extract statistics per
frame of an HDR video. Listing 5.1 shows the main code of the function `hdrvAnalysis.m`, which can be found in the folder `IO_video`. This
function takes an HDR video structure, `hdrv`, as input, and extracts the
minimum (`stats_v(:,1)`), maximum (`stats_v(:,2)`), robust minimum
(`stats_v(:,3)`), robust maximum (`stats_v(:,4)`), mean (`stats_v(:,5)`),
logarithmic mean (`stats_v(:,6)`), and histogram (`hists_v`) of the luminance channel of each frame of the video.

```
name = RemoveExt(filenameOutput);
ext  = fileExtension(filenameOutput);

%create an output video stream
```

```
writerObj = 0;
if(strcmp(ext, 'avi') == 1 | strcmp(ext, 'mp4') == 1)
    writerObj = VideoWriter(filenameOutput, tmo_video_profile);
    writerObj.FrameRate = hdrv.FrameRate;
    writerObj.Quality = tmo_quality;
    open(writerObj);
end
%preprocess the HDR video stream
[hdrv, stats_v, hists_v] = hdrAnalysis(hdrv);
...
```

Listing 5.2. MATLAB code: The first step of a generic video TMO, which prepares the output video stream and analyzes the HDR video stream.

```
%open the HDR video stream
hdrv = hdrvopen(hdrv);
for i=1:hdrv.totalFrames
    %fetch physical values
    [frame, hdrv] = hdrvGetFrame(hdrv, i);
    frame = RemoveSpecials(frame);
    frame(frame < 0) = 0;
    %tone map
    ...
    frameOut = ...
    if(bsRGB) %sRGB encoding
        frameOut = ConvertRGBtosRGB(frameOut, 0);
    else %gamma encoding
        frameOut = GammaTMO(frameOut, tmo_gamma, 0, 0);
    end
    %clamp values in [0,1]
    frameOut = ClampImg(frameOut, 0, 1);
    if(writerObj ~= 0) %store frames as a video
        writeVideo(writerObj, frameOut);
    else %store frames as a sequence of images
        nameOut = [name, sprintf('%.10d',i), '.', ext];
        imwrite(frameOut, nameOut);
    end
end
%close the output video stream
if(writerObj ~= 0)
    close(writerObj);
end
%close the HDR video stram
hdrvclose(hdrv);
```

Listing 5.3. MATLAB code: The second step of a generic video TMO, the main loop in which each each frame is tone mapped and stored.

Often video TMOs have two common steps. The first step comprises opening an LDR video stream, writerObj, for the output (unless the tone mapped video is stored as a series of individual static images), gathering statistics from the HDR video stream, `hdrv`, and running other pre-processing operations (e.g., computing optical flow). Listing 5.2 shows a generic structure of this first part of a video TMO. The second step tone maps each frame of the video (see the loop), and stores each tone mapped frame either inside `writerObj` or as single LDR images; see Listing 5.3. Once all frames are tone mapped, `writerObj` (if open) and `hdrv` are closed.

5.3   Global Temporal Methods

Global video TMOs, as is the case with global TMOs for still images, apply the same tone curve to all pixels of a frame of an HDR video. However, this tone curve can vary from frame to frame because it may adapt to lighting changes. This section introduces the main techniques for adapting the tone curve to avoid global flickering when tone mapping frames. Most of these techniques have been applied to global TMOs for still images, but they can be employed for local TMOs as well without global flicking. However, local flickering may appear because local coherence is not enforced.

Note that some TMOs meant for still images that adapt to illumination changes by modeling the HVS [182,217,302] (see Section 3.2.6) can produce tone mapped HDR videos without global flickering [123]. This is because they take into account the adaptation state of the previous frame.

Filtering statistics temporally.   A solution for reducing global flickering in straightforward way is to compute global statistics over a set of frames. For example, Kang et al. [188] extended Reinhard et al.'s operator [323] to HDR videos based on this concept. The authors decided to compute the logarithmic mean luminance, $L_{\text{w, H},i}$, for the current frame i as

$$L_{\text{w, H},i} = \exp\left(\frac{1}{N \times (\Delta t + 1)} \sum_{k=i-\Delta t}^{i} \sum_{j=1}^{N} \log\left(L_{\text{w},k}(\mathbf{x}_j) + \epsilon\right) \right), \qquad (5.1)$$

where N is the total number of pixels of a frame, Δt is the number of previous frames to take into account, $\mathbf{x}_j$ is the pixel coordinate at the j-th pixel, and $\epsilon > 0$ is a small constant used for avoiding singularities; e.g., $\epsilon = 10^{-6}$. Note that this process reproduces the effect of a low pass filter; see Figure 5.3.

The authors did not investigate how to choose Δt, and setting it to a fixed value does not preserve lighting changes. For example, the *Carousel* presents a moment where the lights switch off (around frame 45 in Figure 5.3(a)), and this change is not preserved due to excessive smoothing.

A possible solution to this problem is to adapt Δt per frame according to the scene luminance as introduced by Ramsey et al. [319]. The authors

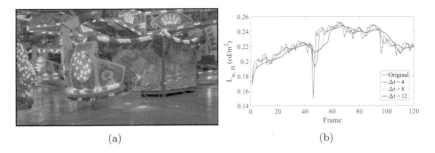

(a)   (b)

Figure 5.3. An example of computing $L_{w, H,i}$ over a set of frames for the *Carousel* HDR video sequence [137]: (a) A tone mapped frame of the sequence. (b) The plot of the original $L_{w, H,i}$ and its filtered versions with different Δt values for Equation (5.1). Note that abrupt changes are smoothed with this approach leading to the reduction of global filtering. (The original HDR video is courtesy of Jan Fröhlich.)

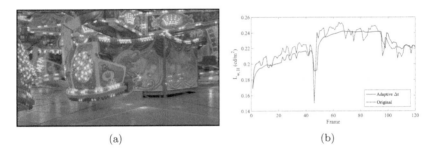

(a)   (b)

Figure 5.4. An example of computing $L_{w, H,i}$ over a set of frames for the *Carousel* HDR video sequence [137]: (a) A tone mapped frame of the sequence. (b) The plot of the original $L_{w, H,i}$ and its filtered version with per frame adaptive Δt value. Note that abrupt changes are filtered but maintained. (The original HDR video is courtesy of Jan Fröhlich.)

allow Δt to vary between 5 (to reduce global flickering) and 60 (to avoid excessive smoothing). The first step of the algorithm is to compute and store the statistic, x, to be filtered per frame for all frames. Subsequently, the filtered statistics for the i-th frame, s_i', is computed by averaging the statistic of previous frames. The averaging stops when either more than 60 frames have been averaged or s_k, the statistic of the k-th previous frame, is not within a tolerance range, i.e., $[s_i \cdot 0.9, s_i \cdot 1.1]$. The authors tested this approach by filtering $L_{w, H}$ by extending Reinhard et al.'s operator [323] for handling HDR videos; see Figure 5.4.

```
%compute the number of frames for smoothing
minLhi = L_h(i) * 0.9;
```

```
maxLhi = L_h(i) * 1.1;
j = i;
while((j > 1) && ((i - j) < 60))
    if((L_h(j) > minLhi) && (L_h(j) < maxLhi))
        j = j - 1;
    else
        if((i - j) < 5)
            j = j - 1;
        else
            break;
        end
    end
end

%compute a smoothed L_ha
L_ha = 0.0;
for k=i:-1:j
    L_ha = L_ha + log(L_h(k));
end
L_ha = exp(L_ha / (i - j + 1));

%fetch the i-th frame
[frame, hdrv] = hdrvGetFrame(hdrv, i);
%only physical values
frame = RemoveSpecials(frame);
frame(frame < 0) = 0;

%tone map the current frame
frameOut = ReinhardTMO(frame, tmo_alpha, tmo_white, 'global',
    -1, L_ha);
```

Listing 5.4. MATLAB code: The inner loop of Ramsey et al. [319] video tone mapper.

Listing 5.4 shows the inner loop of function RamseyTMOv.m, which implements Ramsey et al.'s video operator [319]. The full code may be found in the folder Tmo_video. Before this loop, which tone maps each frame, the function hdrvAnalysis.m is executed in order to calculate the logarithmic mean luminance of all frames and these values are stored into the array L_h. Note that the first part of the inner loop computes the adaptive number of frames for the current frame, (i - j + 1).

Temporal leaky integration. Kiser et al. [196] proposed to use a leaky integrator applied to all global statistics of an operator in order to reduce the global flickering when tone mapping an HDR video. Given a global statistic s_i computed at the i-th frame, its temporally stable value, s'_i, is defined as

$$s'_i = \begin{cases} s_i & \text{if } i = 0, \\ (1 - \alpha) \cdot s'_{i-1} + \alpha \cdot s_i & \text{otherwise,} \end{cases} \qquad (5.2)$$

where α is a time constant for the given statistic whose value is in the range $[0, 1]$, which the authors suggest to set to $\alpha = 0.98$. The authors also tested this approach on Reinhard et al.'s operator [323]. Figure 5.5 shows an example of applying the leaky integration in Equation (5.2).

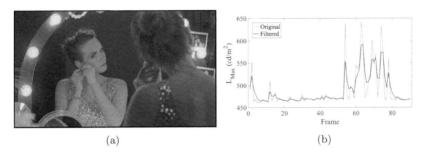

(a)   (b)

Figure 5.5. An example of the leaky integrator method by Kiser et al. [196] applied to the *Showgirl* HDR video sequence [137]: (a) A tone mapped frame of the sequence. (b) The plot of the maximum luminance value, $L_{w, max}$, before (blue plot) and after filtering (red plot). Note that abrupt changes are smoothed with this approach leading to the reduction of global filtering. (The original HDR video is courtesy of Jan Fröhlich.)

```
%fetch the i-th frame
[frame, hdrv] = hdrvGetFrame(hdrv, i);
%only physical values
frame = RemoveSpecials(frame);
frame(frame < 0) = 0;

%compute the maximum luminance
L = lum(frame);
maxL = max(L(:));

if(i == 1)
    maxL_prev = maxL;
end

%leaky integration
maxL_i = (1.0 - alpha) * maxL_prev + alpha * maxL;
maxL_prev = maxL_i;
```

Listing 5.5. MATLAB code: The inner loop of a leaky integrator [196] for smoothing the maximum luminance.

Listing 5.5 shows a fragment of the inner tone mapping loop of Kiser et al.'s method [196]. In this piece of code, the filtered statistic is the maximum luminance of the current frame, maxL. After fetching the current frame and computing its luminance channel, Equation (5.2) is applied in a straightforward way.

Temporal filtering. Van Hateren [387] proposed a new video TMO based on a model of human cones. The operator works in Troland units (Td), which represent the measure of retinal illuminance, I, i.e., the scene luminance in cd/m^2 multiplied by the pupil area in mm^2 (e.g., $10mm^2$).

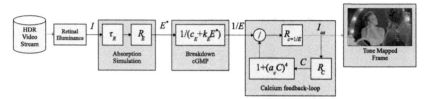

Figure 5.6. The pipeline for cone video tone mapping [387]. Note that τ's boxes are first-order low-pass filters in time. Here $\tau_R = 3ms$, $\tau_E = 8.7ms$, $\tau_C = 3ms$, $c_\beta = 2.8 \cdot 10^{-3}(ms)^{-1}$, $k_\beta = 1.6 \cdot 10^{-4}(ms)^{-1}/Td$, and $a_c = 9 \cdot 10^{-2}$. (The original HDR video is courtesy of Jan Fröhlich.)

The temporal TMO is designed for HDR videos and presents low-pass temporal filters for removing photon and source noise; see Figure 5.6. The TMO starts by simulating the absorption of retinal illuminance by visual pigments, which are modeled by two low-pass temporal filters and described in terms of a differential equation:

$$\frac{dy}{dt} + \frac{1}{\tau}y(t) = \frac{1}{\tau}x(t),$$

where τ is a time constant, and $x(t)$ and $y(t)$ are the input and the output respectively at time t. At this point, a strong non-linear function is applied to the result of low-pass filters E^* for simulating the breakdown of cyclic guanosine monophosphate (cGMP) by enzymes (cGMP is a nucleotide that controls the current across the cell membranes):

$$\alpha = \frac{1}{\beta} = (c_\beta + k_\beta E^*)^{-1},$$

where $k_\beta E^*$ is the light-dependent activity of an enzyme, and c_β is the residual activity. The breakdown of cGMP is counteracted by the production of cGMP, a highly nonlinear feedback loop under control of intercellular calcium. This system is modeled by a filtering loop that outputs the current across the cell membrane, I_{os} (the final tone mapped value), by the outer segment of a cone.

The author applied this TMO and its inverse to uncalibrated HDR movies and images, which were scaled by the logarithmic mean. The results showed that the method does not need gamma correction, removes noise, and presents light adaptation. However, the operator can introduce motion blur in video sequences, and it can only handle content with limited dynamic range (i.e., 10,000:1), otherwise it can cause saturation in very dark and bright regions.

Filtering the tone curve temporally. The display adaptive TMO [242] (see Section 4.1.2) has a different approach to temporal coherency. Instead of filtering statistics that are used by a TMO, it filters the tone curve in the time domain. The authors noted that the sensitivity of the HVS for temporal changes depends on the spatial frequency, which varies from ≈ 0.5Hz to ≈ 4Hz. Therefore, to avoid flickering a low-pass filter with a cut-off frequency of 0.5Hz is applied to the tone curve; see Figure 5.7.

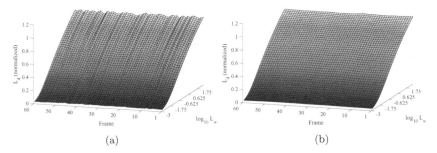

Figure 5.7. An example of filtering the tone curve [242] applied to the *Smith Hammering* HDR video sequence [137]: (a) The plot of a tone curve (Drago et al.'s TMO [116]) evolving over frames. (b) The plot in (a) after filtering.

A general framework. The methods mentioned above have two main disadvantages. The first one, a practical disadvantage, is that the code of the TMO when extended to video sequences require changes. These changes can be minimal, such as adding the filtered statistic as input, but they are required; potentially introducing bugs or multiple versions of the same code. The second disadvantage is that these methods avoid global flickering, but TBI may still occur during tone mapping. To solve both these issues, Boitard et al. [68] proposed to preserve the relative difference of brightness between frames in a straightforward way. Assuming that a TMO for static images has been used to tone map an HDR video per frame, the tone mapped luminance values of the i-th frame, $L_{d,i}$, need to satisfy the following equality:

$$\frac{L_{\text{w, H},i}}{L_{\text{w, H, max}}} = \frac{L_{\text{d, H},i}}{L_{\text{d, H, max}}}, \qquad (5.3)$$

where $L_{w,\,H,i}$ and $L_{d,\,H,i}$ are, respectively, the logarithmic luminance mean of the HDR i-th frame and the tone mapped i-th frame. $L_{w,\,H,\,max}$ and $L_{d,\,H,\,max}$ are the maximum logarithmic luminance mean over the HDR video and the tone mapped video. The idea behind Equation (5.3) is that all frames are anchored to the one which is the brightest, so maintaining the brightness coherency. Equation (5.3) leads to the following scaling:

$$L_{d',i}(\mathbf{x}) = L_{d,i}(\mathbf{x}) \cdot \left(\frac{L_{w,\,H,\,i} \cdot L_{d,\,H,\,max}}{L_{d,\,H,\,i} \cdot L_{w,\,H,\,max}} \right). \tag{5.4}$$

Figure 5.8 shows how Equation (5.4) can preserve the brightness of parts of the scene in HDR videos.

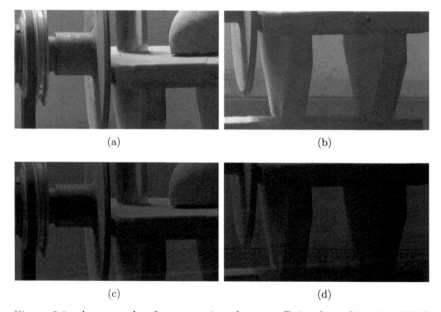

(a)   (b)

(c)   (d)

Figure 5.8. An example of a comparison between Boitard et al.'s video TMO (**B**) [68] and Ramsey et al.'s TMO (**R**) [319] applied to a panning sequence of the Demijohn HDR image: (a) Frame 43 of the sequence tone mapped with **R**. (b) Frame 48 of the sequence tone mapped with **R**. (c) Frame 43 of the sequence tone mapped with **B**. (d) Frame 43 of the sequence tone mapped with **B**. Note that the brightness of the lathe varies in (a) and (b), but it is preserved in (c) and (d).

Boitard et al. [70] extended this method to preserving the local brightness coherency. This is achieved in two steps. Firstly, a histogram-based segmentation is applied to all frames. Then, the logarithmic luminance mean is computed for each segment and each frame. All these values are fed into a histogram that is used to define video zones. Secondly, the scaling in Equation (5.4) is applied to each video zone of each frame. To avoid seam artifacts, spatial blending is employed at the boundaries between zones using Gaussian weights. The main limitation of these methods [68, 70] is that they require analysis of the full HDR video for performing high quality anchoring and segmentation, which makes these solutions unsuitable for live broadcasting. Moreover, the global operator produces videos in which the dynamic range in the final LDR frame is low [123].

5.4 Temporal Local Methods

Local TMOs for HDR videos, as in the case of local TMOs for still images, apply a tone curve that adapts itself to each pixel of a frame of an HDR video. Furthermore, this tone curve can vary from frame to frame to account for lighting changes. In this way, local details are enhanced or highlighted as is the case of local TMOs for still images. A straightforward approach to achieve a local tone mapping for HDR videos is to apply a local TMO to each frame independently using temporally filtered statistics as described in the previous section. Although this approach would not generate global flickering, local flickering may appear due to the fact that local statistics may vary from one frame to another.

This section introduces the main techniques for achieving local TMOs that maintain temporal coherency globally and locally. Most of these techniques rely on spatio-temporal filtering and the use of optical flow.

5.4.1 Temporal Coherence by Energy Minimization

To enforce temporal coherency, Lee and Kim [219] proposed to integrate per-pixel motion information into the Poisson equation of the Fattal et al.'s operator [129] (see Section 3.4.2).

The first step of this operator is to estimate the optical flow, (u_i, v_i), between two successive HDR frames (one at the $i-1$-th frame, and the other at the i-th frame) in the sequence such that:

$$L_{w,i}(x, y) = L_{w,i-1}(x - u_i(x, y), y - v_i(x, y)).$$

Note that several methods can be employed to compute (u_i, v_i) such as block matching; see Szeliski [361] for an overview on the topic.

The second step is to tone map all frames of the video exploiting optical flow. The first frame of the sequence, $L_{d,0}$, is tone mapped using the Fattal et al.'s TMO for still images. To maintain temporal coherence between frames, the main idea is to warp the previous tone mapped, $L_{d,i-1}$, using

(u_i, v_i) and to add it as a smoothing constraint when solving Fattal et al.'s minimization function. This leads to the following energy:

$$\arg \min_{I_i} \sum_{x,y} \left\| \nabla I_i(x,y) - \mathbf{G}(x,y) \right\|^2 + \lambda \sum_{x,y} \left(I_i(x,y) - \log warp[L_{\mathrm{d},i-1}](x,y) \right)^2,$$

(5.5)

where $I_i = \log L_{\mathrm{d},i}$, $warp[\cdot]$ is the warping function that uses the optical flow (u_i, v_i) at the i-th frame, and λ is a regularisation term, which authors suggest to set to 0.3 after their experiments.

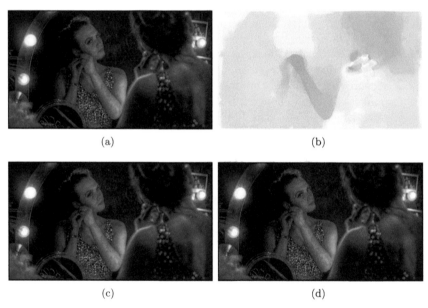

(a)   (b)

(c)   (d)

Figure 5.9. An example of Lee and Kim's operator [219] applied to the *Showgirl* HDR video sequence [137]: (a) A frame tone mapped at previous step with the operator. (b) The optical flow between the HDR version of the frame in (a) and its next frame [358]; blue-green is low motion, and yellow-red is high motion. (c) The frame in (a) warped using the optical flow in (b); to be used in Equation (5.5). (d) The result of the minimization in Equation (5.5) for generating the next frame. (The original HDR video is courtesy of Jan Fröhlich.)

This method can achieve temporally coherent tone mapped videos by enforcing coherence with previous frame as a minimization constraint; see Figure 5.9. However, it can produce ghosting artifacts [123] when optical flow is not accurate. Moreover, noise can be enhanced due to the nature of the algorithm that boosts the intensity of small gradients.

A more general framework for blind temporal consistency was introduced by Bonneel et al. [72]. This framework, as with Lee and Kim's [219], enforces temporal consistency using backward optical flow. The main differences are two-fold: all frames are tone mapped with a TMO of choice for still images at the beginning of computations, and the introduction of a novel energy function is introduced. This energy function to minimize is defined as

$$\arg\min_{L_{d,i}} \sum_{x,y} \left\| \nabla L_{d,i}(x,y) - \nabla L_{d,i}^s(x,y) \right\|^2 +$$

$$\omega(x,y) \left\| L_{d,i}(x,y) - warp[L_{d,i-1}](x,y) \right\|^2, \quad (5.6)$$

where $L_{d,i}^s$ is the tone mapped i-th frame using a TMO for still images. $L_{d,0} = L_{d,0}^s$ (the boundary condition), and $\omega(x,y)$ is a weighting function that is defined as

$$\omega(x,y) = \lambda \exp\left(-\alpha \| L_{w,i}(x,y) - warp[L_{w,i-1}](x,y)\|^2\right), \quad (5.7)$$

where λ is a parameter that controls the regularization strength, and α is another user parameter which determines the confidence in the warping function. Note that Equation (5.7) imposes temporal consistency only when the input video is consistent. This is a key feature in the approach. Equation (5.6) leads to the screened Poisson equation (a standard Poisson equation with a 0th-order term added), which can be solved with numerical methods [60].

Bonneel et al.'s framework is very general. It can be applied to any TMO for still images and not only to Fattal et al.'s operator [129]. Furthermore, it has been applied to different imaging problems for videos such as shading and albedo decomposition, grading, depth prediction, de-hazing, stylization, etc. However, the quality of the results critically depends on high-quality optical flow. The authors highlighted that the use of Patch-Match [53] for dense correspondences (which is computationally faster than many optical flow algorithms) can produce reasonable quality results.

5.4.2   Spatio-Temporal Filtering Approaches

Another approach to achieve local tone mapping for HDR videos is to employ novel filtering strategies. A simple solution could be to apply methods used for the enhancement of LDR videos; e.g., Bennett and McMillan's framework [57] for enhancing low-light LDR videos. The key idea of this framework is a modified spatio-temporal bilateral filter; called ASTA. This filter combines a 2D spatial bilateral filter and an 1D temporal bilateral filter depending on the weight of their normalization term. Eilertsen et al. [123] tried to apply this approach to HDR videos, but their experiments highlighted that it produces several artifacts in HDR video sequences. However, this may be due to the fact that main formulation of the framework is meant for static cameras, and the handling of fully dynamic scenes (both moving camera and people/objects) is not taken into account. Therefore, the need for HDR dedicated approaches is crucial in order to deliver high quality local video tone mapping.

OPL filter. One of the first attempts to achieve a local video TMO was by Benoit et al. [58]. The authors introduced a video TMO, which is based on a biological model of the HVS. The operator employs a sigmoid function for range reduction, which takes as input a spatio-temporal adaptation. This adaptation is computed by modeling the cellular interactions of the Outer Plexiform Layer (OPL) in the retina as a spatio-temporal high-pass filter. However, this method may produce high frequency flickering and ghosting artifacts [123], possibly due to the fact that the filtering does not track the content.

Permeability filter. Similarly to Durand and Dorsey's operator [118] (see Section 3.4.1), Adyın et al. [32] proposed a TMO for HDR videos that decomposes an HDR frame into base and detail layers. In order to achieve global and local temporal coherency and avoid ghosting artifacts, the authors introduced two novel filters, which are based on iterative applications of shift-variant filters. The first filter is spatio-temporal and it is employed for computing a temporal coherent base layer. The second filter is temporal only, and computes the detail layer such that temporal artifacts with low amplitude and high temporal frequency are minimized. Note that frames are aligned (warped) during the spatio-temporal and temporal processing using both backward and forward optical flows for each frame. The novelty of the approach is the introduction of the *permeability map* and the *flow confidence map* (i.e., quality of the optical flow), which respectively control the spatial and temporal diffusion between neighboring pixels. Once the base and detail layers are computed, the dynamic range of the base can be compressed by applying a tone curve. For example, the authors compressed the base in their experiments using both Durand and Dorsey's compression factor in the logarithmic domain [118] and Drago et al.'s tone curve [116]. Furthermore, this method was tested by running a subjective study against video TMOs tested by Eilertsen et al. [123]. This study highlighted that the method can produce high local contrast with a similar level of temporal coherence as global video TMOs.

Noise-aware video tone mapping. High-end cameras offer noise levels below the perceivable contrast of the HVS. However, after tone-mapping, the noise is boosted and made visible in the tone mapped video. To control noise visibility and to have global and local temporal coherency in real-time,

Eilersten et al. [121] proposed a computationally fast display adaptive video TMO.

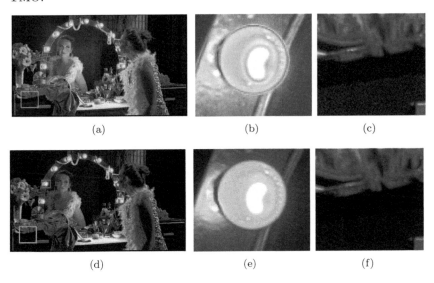

(a)   (b)   (c)

(d)   (e)   (f)

Figure 5.10. An example of Eilertsen et al.'s operator [121] (**RTN**) compared to a naïve approach based on the bilateral filter (**N**) applied to the *Showgirl* HDR video sequence [137]: (a) A frame of the sequence tone mapped with **N**. (b) A scaled-in image of the red box in (a); note the ringing artifacts around the lightbulb. (c) A scaled-in image of the green box in (a); note the noise in the dark areas of the image. (d) The same frame in (a) tone mapped with **RTN**. (e) A scaled-in image of the red box in (d); note the ringing artifacts in (b) are not present using **RTN**. (f) A scaled-in image of the green box in (d); note that noise has been removed compared to (c). (The original HDR video is courtesy of Jan Fröhlich. Images were generated with the help of Gabriel Eilertsen.)

The operator is based on base and detail layer decomposition (see Section 3.4.1) and it is composed of three steps. The first one is a tone mapping curve for reducing the dynamic range that is a modified version of Mantiuk el al.'s tone curve [242] in which a noise model is introduced to avoid the amplification of noise. Furthermore, to achieve real-time performance, the minimization process for computing the tone curve employs a simple contrast distortion measure, the L_2 norm, instead of a model of the HVS as in Mantiuk et al.'s work. Temporal coherence is enforced by filtering the tone cure; see Section 5.3. The second step is an iterative and computationally fast spatial edge-aware filter inspired by both the bilateral filter [349, 368] and the anisotropic diffusion [307]. This filter is designed to extract base and detail layers avoiding the extraction of false edges (or rings) around soft edges exploiting the Tukey's biweight function [118] and robust gradient formulation. The third step comprises further processing of the detail layer that is modulated relative to the visibility of noise in the tone mapped frame. Note that only the first step presents an explicit mechanism for enforcing temporal coherence. However, the next two steps are designed to avoid noise enhancement, which can cause local flickering in dark areas, and to extract stable edges across the frames.

The authors ran a subjective study showing that their method has barely visible artifacts compared to other methods; see Figure 5.10. Compared to Adyın et al. [32], this method can produce higher quality tone reproduction when tone mapping complicated image transitions because the optical flow employed in Adyın et al.'s work [32] may not work well. In addition, the method can be implemented in real-time, with an average of 21.5 ms per frame, on graphics hardware at HD resolutions.

5.5   Real-Time HDR Video Tone Mapping

When tone mapping HDR videos, an important issue is the computational performance of the method. This is because real-time tone mapping of HDR video content for an LDR display is essential for applications, such as live broadcasting, video games etc. A key problem is that many TMOs require computationally expensive image processing operations such as linear filtering, edge-aware filtering, numerous color spaces conversions, wavelet decomposition, etc.

Ad-hoc solutions. The first approaches to achieve real-time tone mapping used the graphics hardware of the time [97, 145, 333, 340]. As the performance of graphics hardware improved, it became possible for global operators to be mapped onto a single shader [41] and the statistics to be computed either using a map-reduction paradigm or shared global counters. Furthermore, computationally expensive filtering operations for local operators can now be computed by either exploiting languages for the imaging domain [314] or using computationally efficient approximations [15,44,89]. Of course re-implementing TMOs for the latest version on any GPU lends itself to the introduction of bugs and thus thorough testing is always required.

General frameworks. To overcome the limited flexibility offered by implementing directly on GPU's and FPGA's, Artusi et al. [21,22,25] proposed two frameworks that exploit the possibility to use any state-of-the-art global and local operators.

The first framework [22] allows an efficient subdivision of the workload between the CPU and the GPU. While the GPU resolves the computationally simple but costly stages of the algorithm, the CPU applies the tone mapping using an existing implementation of the TMO. Such an approach does not require modifications of the rendering pipeline or of the original TMO.

The second framework [21, 25] exploits a straightforward and efficient idea. Based on the fact that global and local TMOs have different characteristics as described in Chapter 3, the authors proposed an algorithm that combines the best of both approaches. The local tone mapping is performed only in high frequency image regions where the visibility of details can be an issue. In low frequency regions, global tone mapping is employed to save computational resources without degrading quality.

5.6 Summary

Despite the significant advances in HDR technology and video standardization (ISO/IEC/SC29 WG 11 – MPEG), tone mapping in video applications is still required.

Operator	GFA	LFA	TBI	TOI	THI	G	TN
Pattanaik et al. [302]	Yes	N/A	No	No	No	N/A	No
Kang et al. [188]	Yes	N/A	No	No	No	N/A	No
Ramsey et al. [319]	Yes	N/A	Yes	No	No	N/A	No
Irawan et al. [182]	Yes	N/A	No	No	No	N/A	No
Ledda et al. [302]	Yes	No	No	No	No	N/A	No
Van Hateren [387]	Yes	N/A	No	No	No	N/A	Yes°
Lee and Kim [219]	Yes	Yes	No	No	No	No	No
Mantiuk et al. [242]	Yes	N/A	Yes	No	No	N/A	No
Benoit et al. [58]	Yes	Yes	No	No	No	N/A	Yes°
Kiser et al. [196]	Yes	No	No	No	No	N/A	No
Boitard et al. [68]	Yes	No	Yes	Yes	Yes	N/A	No
Boitard et al. [70]	Yes	Yes	Yes	Yes	Yes	N/A	No
Adyın et al. [32]	Yes	Yes	Yes	No	No	Yes	Yes
Bonneel et al. [72]	Yes	Yes	Yes	No	No	No	Yes*
Eilertsen et al. [121]	Yes	Yes	Yes	No	No	N/A	Yes

Table 5.1. This table summarizes different video TMOs (or TMOs for still images that work in the temporal domain) and their ability to reduce artifacts (see Section 5.1). If an operator (in its full form as described in the original paper) does not introduce an artifact, the cell will have an *N/A*. The superscript * means that it depends on the input TMO. The superscript ° means that the method may introduce motion blur when dealing with temporal noise.

This is because, although improved, display technology with its higher dynamic range and peak luminance, is often still not sufficient to accommodate the full dynamic range present in some HDR video footage. In addition, backward compatibility is still an important requirement in many applications and for industry this helps limit their investment in integrating new HDR standards into their existing LDR standard ecosystem.

An overview of all methods discussed is presented in Table 5.1. This shows which artifacts the TMO can and cannot deal with. From this table, it is clear that no HDR video TMO, either global or local, can overcome all artifacts yet. Even the most promising operator (in terms of both quality and computational speed), Eilertsen et al. [121], does not preserve TOI. Although TOI is desirable in some circumstances, it typically reduces the dynamic range of the tone mapped video [123]. HDR video tone mapping is a very recent field of endeavor and more research is still needed in order to adequately address remaining challenges, such as temporal noise, computational efficiency, ghost-free frames, and brightness/hue preservation.

6
Expansion Operators for Low Dynamic Range Content

The quick rise in popularity of HDR has meant that the large amount of legacy content is not particularly suited for the new HDR technologies. This chapter presents a number of methods that generate HDR content from LDR. The expansion of LDR content is achieved by transforming LDR content into HDR content using operators commonly termed expansion operator (EO); see Figure 6.1. This expansion allows the large amounts

(a)

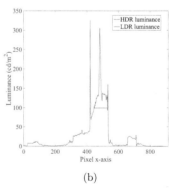

(b)

Figure 6.1. The general concept of expansion operators: (a) A single exposure image. (b) A graph illustrating the luminance scanline at coordinate $y = 448$ (green line in (a)) of this single exposure image. The red line shows the clamped luminance values due to the single exposure capture of the image. The blue line shows the full luminance values when capturing multiple exposures. An expansion operator tries to recover the blue profile starting from the red profile.

of legacy LDR content to be enhanced for viewing on HDR displays and to be used in HDR applications such as image-based lighting. An analogy to expansion is the colorization of legacy black and white content for color displays [222, 231, 415].

An EO is defined as a general operator g over an LDR image I as

$$g(I) = \mathbb{D}_i^{w \times h \times c} \rightarrow \mathbb{D}_o^{w \times h \times c},$$

where $\mathbb{D}_i \subset \mathbb{D}_o$, and w, h, and c are, respectively, the width, the height, and the number of color channels of I. In the case of LDR images $\mathbb{D}_i = [0, 255]$, and $\mathbb{D}_o$ is the expanded domain; in the case of single precision floating point $\mathbb{D}_o = [0, 3.4 \times 10^{38}] \subset \mathbb{R}$.

This is an ill-posed problem since information is missing in over-exposed and under-exposed regions of the image/frame, see Figure 6.1. Typically, an EO has the followings steps:

1. Linearization creates a linear relationship between real-world radiance values and recorded pixels.

2. Expansion of pixel values increases the dynamic range of the image. Usually, low values are compressed, high values are expanded, and mid values are kept as in the original image.

3. Over/under-exposed reconstruction generates the missing content in the over-exposed or under-exposed areas of an LDR image.

4. Artifacts reduction decreases artifacts due to quantization or image compression (i.e., JPEG and MPEG artifacts), which can be visible after the expansion of pixel values.

5. Color correction keeps colors as in the original LDR image. During expansion of pixel values colors can be desaturated. This is the opposite problem of what happens in tone mapping.

Researchers have also proposed an optional color-correction step before linearization [158, 423]. In this step, the luma and chroma of clipped pixels are enhanced; typically exploiting spatial coherency and well-exposed pixels near over-exposed ones. The output of these algorithms are either LDR images [158, 253, 423] or images with a moderate dynamic range [29, 337]. These algorithms should not be confused with EO ones. Typically, these algorithms greatly enhance images when over-exposed areas are relatively small, i.e., 1–5% of the image; see Figure 6.2. Otherwise, they can produce artifacts such as a biased colors [158].

As in the case of TMOs, color gamut is changed during expansion obtaining desaturated colors. To solve this problem, a straightforward technique such as color exponentiation can be applied; see Equation (3.2). Here

(a)   (b)

Figure 6.2. An example of the color correction technique by Assefa et al. [29]:
(a) The input LDR image. (b) The application of the color-correction algorithm.
Note that details in the petals are fully recovered. The exposure values in both
images are set to f-stop -1 for the purpose of visualization. (Tania Pouli helped
in the generation of the output result.)

the exponent s needs to be set in the interval $(1, +\infty)$ (as opposed to $(0, 1]$)
to increase the saturation. However, color management for content expansion has not been exhaustively studied. One of the first studies [1] based
on a subjective evaluation, on this matter showed that expansion methods
combined with chromatic adaptation transforms give more reliable results.
This study also suggests that more research needs to be carried out on this
topic.

EOs are frequently referred to in the literature as inverse tone mapping
operators (iTMOs) [48] or reverse tone mapping operators (rTMOs) [324].
However, neither of these two terminologies is strictly correct, and they do
not fully capture the complex operations required to convert LDR content
into HDR one. In this work the nomenclature of iTMOs being EOs that
invert TMOs is adopted. However, many TMOs are not invertible because
they require the inversion of a partial differential equation (PDE), or other
non-invertible operations. Examples of these TMOs are LCIS [374], the
fast bilateral filter TMO [118], gradient domain compression TMO [129],
etc. rTMOs, on the other hand, only refer to those EOs that reverse the
order of TMO operations.

6.1   EO MATLAB Framework

Often EOs, independently to which category they belong, have two common steps. This section describes the common routines that are used by
most, but not all, EOs. The first one is the linearization of the signal and
extraction of the luminance channel. Linearization is achieved by applying

gamma removal. This works in most cases, such as DVDs and TV signals with a sRGB profile. It is a straightforward and computationally fast method that can work in most cases. Other methods can be applied and these are discussed in Section 6.2. Note that EOs typically work on the luminance channel to avoid color expansion. The second step is the restoration of color information in the expanded image. The implementation of these two steps is shown in Listing 6.1 and Listing 6.2.

In the first step, shown in Listing 6.1, the input image img is checked to see whether it is a three-color or a grayscale image using the function check13Color.m under the util folder. Then, input parameters are verified if they are provided; if not, they are set equal to default parameters that are suggested by the authors in their original work. An input parameter common to all EOs is gammaRemoval. If it is higher than zero this means that img has not been gamma corrected and an inverse gamma correction step is computed exponentially by applying the gammaRemoval parameter to the input image img. Otherwise, img is already corrected. At this point, luminance is extracted using the lum function applied to the whole input image img.

In the last step of Listing 6.2, ChangeLuminance.m in the folder Tmo/util is applied to img for removing the old luminance, L, and substituting it with Lexp, obtaining imgOut.

```
check13Color(img);
if(exist('maxOutput', 'var') | ~exist('gammaRemoval', 'var'))
    maxOutput = -1;
    gammaRemoval = -1;
end
%set maxOuput to a default value
if(maxOutput < 0.0)
    maxOutput = 3000.0;
end
%remove gamma correction
if(gammaRemoval > 0.0)
    img = img.^gammaRemoval;
end
%compute luminance channel
L = lum(img);
```

Listing 6.1. MATLAB code: The initial steps common to all EOs.

```
imgOut = ChangeLuminance(img, L, Lexp);
```

Listing 6.2. MATLAB code: The final steps common to all EOs.

An optional step, as with tone mapping, is color correction. ColorCorrection.m, ColorCorrectionLinear.m, and ColorCorrection

Pouli.m under the ColorCorrection folder can be applied to increase the saturation. In the case of using either ColorCorrection.m or Color CorrectionLinear.m, cc_factor needs to be manually set in the range $(1, +\infty)$. For ColorCorrectionPouli.m, the LDR image is assigned to imgHDR and the expanded image is assigned to imgTMO. Note that not all EOs' implementations are provided in this chapter because some of them are very complex systems and composed of many parts making them difficult to describe in their entirety.

6.2   Linearization of the Signal Using a Single Image

Linearization of the signal is a very important step in many EOs since working in a linear space means that there is more control and predictability over the expansion. In an unknown space, predicting how an expansion will behave can be difficult. Figure 6.3 shows a typical example of this problem. Moreover, precise/estimated measurements of scene radiance are needed in order to recover statistical data of the scene such as mean, geometric mean, standard deviation, etc.

A standard straightforward solution to linearization, as described in Section 6.1, involves the use of an inverse gamma, particularly if the gamma is known as is the case of TV and DVD signals where gamma has a value of 2.2. For other cases, access or knowledge of the device, which captured the content to expand, can improve the efficiency of the linearization step. In this case, the linearization is performed through the estimation of the CRF of the acquisition device. This function can be calculated using methods presented in Section 2.1.1. Furthermore, RAW data capturing can avoid this step because RAWs are stored linearly. However, the most difficult case is faced when the CRF of the digital camera is not available. The next subsections describe various solutions to estimate the CRF from a single frame/image.

6.2.1   Blind Inverse Gamma Function

Computer-generated images or processed RAW photographs are usually stored with gamma correction. Farid [127] proposed an algorithm based on multispectral analysis to blindly linearize the signal when gamma correction has been applied to an image. He observed that gamma correction adds new harmonics into a signal. For example, given a signal $y(x) = a_1 \sin(\omega_1 x) + a_2 \sin(\omega_2 x)$, if $y(x)$ is gamma corrected and approximated using the Taylor

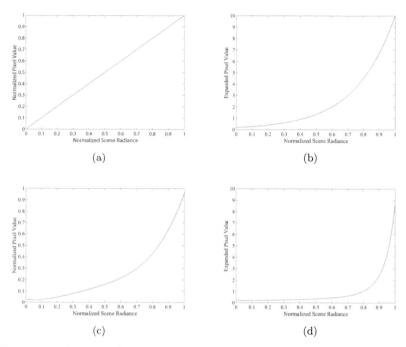

Figure 6.3. An example of the need of working in linear space: (a) Measured values from a sensor in the linear domain. (b) The expansion of the signal in (a) using $f(x) = 10e^{4x-4}$. (c) Measured values from a sensor with an unknown CRF applied to it. (d) The expansion of signal (c) using f. Note that the expanded signal in (d) has a different shape from (b), which was the desired one.

expansion second term, $y(x)_2^\gamma$, the result is

$$
\begin{aligned}
y(x)^\gamma \approx y(x)_2^\gamma =\, & a_1 \sin(\omega_1 x) + a_2 \sin(\omega_2 x) \\
& + \frac{1}{2}a_1^2(1 + \sin(2\omega_1 x)) + \frac{1}{2}a_2^2(1 + \sin(2\omega_2 x)) \\
& + 2a_1 a_2(\sin(\omega_1 + \omega_2) + \sin(\omega_1 - \omega_2)).
\end{aligned}
$$

As can be seen, new harmonics are introduced after gamma correction that are correlated with the original ones: $2\omega_1$, $2\omega_2$, $\omega_1 + \omega_2$, and $\omega_1 - \omega_2$.

High-order correlations introduced by the gamma in a one-dimensional signal $y(x)$ can be estimated using high-order spectra. For example, a normalized bi-spectrum correlation estimates third order correlations [259]:

$$
b^2(\omega_1, \omega_2) = \frac{E[Y(\omega_1)Y(\omega_2)\overline{Y}(\omega_1 + \omega_2)]}{E[|Y(\omega_1)Y(\omega_2)|^2]E[|Y(\omega_1 + \omega_2)|^2]},
$$

where Y is the Fourier transform of y. The value of $b^2(\omega_1, \omega_2)$ can be estimated using overlapping segments of the original signal y:

$$\hat{b}(\omega_1, \omega_2) = \frac{|\frac{1}{N} \sum_{k=0}^{N-1} Y_k(\omega_1) Y_k(\omega_2) \overline{Y}_k(\omega_1 + \omega_2)|}{\sqrt{\frac{1}{N} \sum_{k=0}^{N-1} |Y_k(\omega_1) Y_k(\omega_2)|^2 \frac{1}{N} \sum_{k=0}^{N-1} |Y_k(\omega_1 + \omega_2)|^2}},$$

where Y_k is the Fourier transform of the k-th segment, and N is the number of segments. Finally, the gamma of an image is estimated by applying a range of inverse gamma to the gamma-corrected image and choosing the value that minimizes the measure of the third-order correlations as

$$\epsilon = \sum_{\omega_1 = -\pi}^{\pi} \sum_{\omega_2 = -\pi}^{\pi} |\hat{b}(\omega_1, \omega_2)|.$$

This technique was evaluated using fractal and natural images and showed that a recovered γ can have an average error between 5.3% and 7.5%.

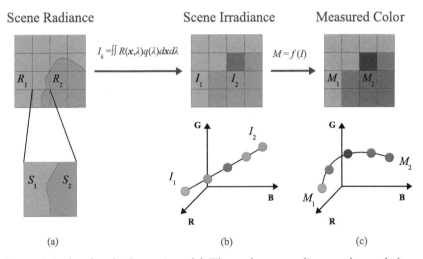

Figure 6.4. A colored edge region: (a) The real scene radiance value and shape at the edge with two radiance values R_1 and R_2. (b) The irradiance pixels, which are recorded by the camera with interpolated values between I_1 and I_2. Note that colors are linearly mapped in the RGB color space. (c) The measured pixels after the CRF application. In this case, colors are mapped on a curve in the RGB color space.

6.2.2 Radiometric Calibration from a Single Image

A general approach for linearization using a single image was proposed by Lin et al. [227]. This approach approximates the CRF by exploiting

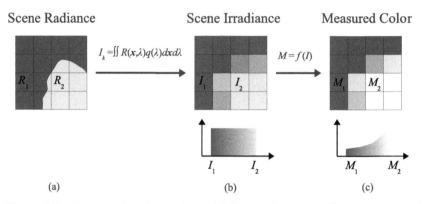

Figure 6.5. A grayscale edge region: (a) The real scene radiance values and shape at the edge with two radiance values R_1 and R_2. (b) The irradiance pixels, which are recorded by the camera. Irradiance values are uniformly distributed. This results in a uniform histogram. (c) The measured pixels after the CRF application. Note that the resulting histogram is non-uniform.

information from edges in the image.

In an edge region Ω there are mainly two colors, I_1 and I_2, separated by the edge; see Figure 6.4(a). Colors between I_1 and I_2 are a linear interpolation of these two and form a line in the RGB color space; see Figure 6.4(b). When the CRF is applied to these values, $M = f(I)$, since it is a general nonlinear function, the line of colors is transformed into a curve (see Figure 6.4(c)). In order to linearize the signal the inverse CRF, $g = f^{-1}$, has to map measured values $M(\mathbf{x})$ to a line defined by $g(M_1)$ and $g(M_2)$. This is solved by finding a g where for each pixel $M(\mathbf{x})$ in the region, Ω minimizes distance from $g(M(\mathbf{x}))$ to the line $\overline{g(M_1)g(M_2)}$:

$$D(g, \Omega) = \sum_{\mathbf{x} \in \Omega} \frac{|(g(M_1) - g(M_2)) \times (g(M(\mathbf{x})) - g(M_2))|}{|g(M_1) - g(M_2)|}. \qquad (6.1)$$

Edge regions, which are suitable for the algorithm, are chosen from non-overlapping 15×15 windows with two different uniform colors along the edge, which is detected using a Canny filter [144]. The uniformity of two colors in a region is calculated using variance. To improve the quality and complete the missing parts of g in Equation (6.1), a Bayesian estimation step is added using inverse CRFs from real cameras [153].

Lin and Zhang [228] generalized Lin et al.'s method [227] to handle grayscale images for which their method cannot be applied. g is now estimated using a function that maps non-uniform histograms of M into uniform ones; see Figure 6.5. They proposed a measure for determining the uniformity of a histogram H of an image, which is defined as

$$N(H) = \frac{1}{b} \sum_{k=I_{\min}}^{I_{\max}} \left[|H(k)| - \frac{|H(k)|}{b} \right]^2 + \frac{\beta}{3} \sum_{n=1}^{3} \left[|H_n| - \frac{|H|}{3} \right]^2,$$

where $|H|$ is the number of pixels in H, $|H(k)|$ is the number of pixels of intensity k, b is the number of grayscale levels, β is an empirical parameter, $I_{\min}$ and $I_{\max}$ are the minimum and maximum intensity in H, and H_n is a cumulative function defined as

$$H_n = \sum_{i=I_{\min}+\frac{(n-1)b}{3}}^{I_{\min}+\frac{nb}{3}-1} |H(i)|.$$

The inverse CRF is calculated similarly to Lin et al.'s method [227], where g is a function that minimizes the histogram uniformity for each edge region:

$$D(g, \Omega) = \sum_{H \in \Omega} \omega_H N(g(H)) \quad \omega_H = \frac{|H|}{b},$$

where ω_H is a weight for giving more importance to a dense histogram. Regions are chosen as in Lin et al. [227], and g is refined as in the previous work. The method can be applied to colored images by applying it to each color channel.

6.3 Decontouring Models for High Contrast Displays

Daly and Feng [101, 102] proposed two methods for extending the bit depth of classic 8-bit images and videos (effectively 6-bit due to MPEG-2 compression) for 10-bit monitors. New LCD monitors present higher contrast, typically around 1,000:1, and a high luminance peak that is usually around 400 cd/m^2. This means that displaying 8-bit data, without any refinement, would entail having the content linearly expanded for higher contrast, resulting in artifacts such as banding and contouring. The goal of Daly and Feng's methods is to create a medium dynamic range image, removing contouring in the transition areas, without particular emphasis on over-exposed and under-exposed areas.

6.3.1 Amplitude Dithering

The first method proposed by Daly and Feng [101], is based on amplitude dithering by Roberts [330]. Amplitude dithering, or noise modulation, is

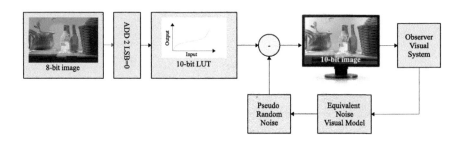

Figure 6.6. The pipeline for bit depth extension using amplitude dithering by Daly and Feng [101].

a dithering technique that simply adds a noise pattern to an image before quantization. This noise pattern is removed when the image needs to be visualized. The bit depth is perceived as higher than the real one because there is a subsequent averaging happening in the display and in the HVS. Roberts' technique was modified to apply it to high contrast displays by Daly and Feng; see Figure 6.6. Subtractive noise was employed instead of additive, since during visualization a monitor can not remove it. The authors modeled the noise by combining the effect of fixed pattern display noise and the one perceived by the HVS, making the noise invisible. They used a CSF, which is a two-dimensional, anisotropic function derived by psychophysical experiments [100]. This CSF is extended in the temporal dimension [411] for moving images, which allows the noise to have a higher variance. Furthermore, it can be shown that the range can be extended by an extra bit.

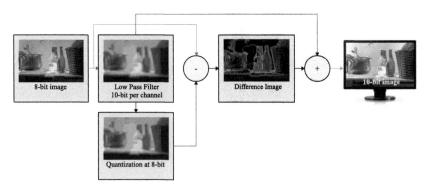

Figure 6.7. The pipeline for bit depth extension using decontouring by Daly and Feng [102].

6.3.2 Contouring Removal

The second algorithm proposed by Daly and Feng [102] presented a different approach. In this, contours are removed instead of being masked with invisible noise. The first step of the algorithm is to filter the starting image at p bits using a low-pass filter; see Figure 6.7. The filter needs to be wide enough to span across the false contours. Note that this operation increases the bit depth to $n > p$ because, during averaging, a higher precision is needed than the one for the original values. Then, this image is quantized at p bits, where any contour that appears is a false one because the image has no high frequencies. Subsequently, the false contours are subtracted from the original image and the filtered image at p bits is added to restore low frequency components. The main limitation of the algorithm is that it does not remove artifacts at high frequencies, but they are hard to detect by the HVS anyway due to frequency masking [131].

6.4 Global Models

Global models are those that apply the same single global expansion function on the LDR content at each pixel in the entire image.

6.4.1 A Power Function Model for Range Expansion

One of the first expansion methods was proposed by Landis [213]. This expansion method, used primarily for relighting digital three-dimensional models from images (see Chapter 7), is based on power functions. The luminance expansion is defined as

$$L_w(\mathbf{x}) = \begin{cases} (1-k)L_d(\mathbf{x}) + kL_{w,\,max}L_d(\mathbf{x}) & \text{if } L_d(\mathbf{x}) \geq R, \\ L_d(\mathbf{x}) & \text{otherwise;} \end{cases}$$

$$k = \left(\frac{L_d(\mathbf{x}) - R}{1 - R} \right)^\alpha, \tag{6.2}$$

where R is the threshold for expansion (which is equal to 0.5 in the original work), $L_{w,\,max}$ is the maximum luminance that the user needs for the expanded image, and α is the exponent of fall-off that controls the stretching of the tone curve.

While this technique produces suitable HDR environment maps for IBL, see Figure 6.8, it may not produce high quality images/videos that can be visualized on HDR monitors. This is because the thresholding can create abrupt changes that result in unpleasant images.

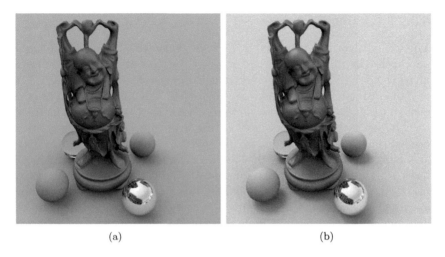

(a)   (b)

Figure 6.8. An example of IBL using Landis' operator [213]: (a) The Happy Buddha is relit using an LDR environment map. (b) The Happy Buddha is relit using the expanded environment map used in (a). Note that directional shadows from the sun are now visible. (The Happy Buddha model is courtesy of the Stanford 3D Models Repository.)

The MATLAB code for Landis' EO [213] is available in the file LandisEO.m under the folder EO. Landis' method takes as input the following parameters: α is l_alpha, the threshold R is l_threshold, and $L_{w, \max}$ is maxOutput.

The code of Landis' EO is shown in Listing 6.3. After the initial steps common to all EOs are performed, the mean luminance value of img is computed using the MATLAB function mean.

```
if(l_threshold <= 0) %expansion from the mean value
    l_threshold = mean(L(:));
end
%find pixels needed to be expanded
toExpand = find(L >= l_threshold);
%expansion using a power function
w = ((L(toExpand) - l_threshold) / (max(L(:)) - l_threshold)).^
    l_alpha;
Lexp = L;
Lexp(toExpand) = L(toExpand) .* (1 - w) + maxOutput * L(
    toExpand) .* w;
```

Listing 6.3. MATLAB code: The expansion method by Landis [213].

Next, the pixels above l_threshold are selected using the MATLAB function find.m and are stored into the variable toExpand. The first line of the expansion step computes the parameter k of Equation (6.2), and the second line directly applies it. Note that only pixels above l_threshold are modified, following the second condition in Equation (6.2). Otherwise, the original luminance is assigned to the variable Lexp.

6.4.2 Linear Scaling for HDR Monitors

In order to investigate how well LDR content is supported by HDR displays, Akyüz et al. [17] ran a series of subjective experiments. The experiments were run to evaluate tone mapped images, single exposure images, and HDR images using a Brightside DR-37p HDR monitor. The experiments involved 22 naïve participants between 20 and 40 years old. In all experiments, ten HDR images ranging from outdoor to indoor, from dim to very bright light conditions, were used. The HDR images had around five orders of magnitude of luminance in order to be mapped directly to the Brightside DR-37p HDR monitor [76].

The first experiment was a comparison between HDR and LDR images produced using various TMOs [118, 215, 323], an automatic exposure (that minimizes the number of over/under-exposed pixels), and an exposure chosen by subjects in a pilot study. Images were displayed on the DR-37p, using calibrated HDR images and LDR images calibrated to match the appearance on a Dell UltraSharp 2007FP 20.1″ LCD monitor. Subjects had the task of ranking images that looked best to them. For each original test image, a participant was required to watch a trial image for two seconds, which was randomly chosen from the different types of images. The experimental results showed that participants preferred HDR images. The authors did not find a significant difference in participant preference between tone mapped and single exposure (automatic and chosen by the pilot) images.

In the second experiment, the authors compared expanded single exposure with HDR and single exposure images (automatic and chosen by the pilot). To expand the single exposure images, they employed the following expansion method in Equation (6.3):

$$L_{\mathrm{w}}(\mathbf{x}) = k \left(\frac{L_{\mathrm{d}}(\mathbf{x}) - L_{\mathrm{d,\,min}}}{L_{\mathrm{d,\,max}} - L_{\mathrm{d,\,min}}} \right)^{\gamma}, \qquad (6.3)$$

where k is the maximum luminance intensity of the HDR display, and γ is the nonlinear scaling factor. For this experiment, images with different γ values equal to 1, 2.2, and 0.45 were generated. The setup and the ranking task were the same as the first experiment. The results showed that brighter chosen exposure expanded images were preferred to HDR images, and vice versa when they had the same mean luminance. Akyüz et al. suggested that mean luminance is preferable to contrast. Finally, another important result is that linear scaling, where $\gamma = 1$, generated the most favored images, suggesting that a linear scaling may be enough for an HDR experience.

```
%compute image statistics
L_max = max(L(:));
L_min = min(L(:));
Lexp = maxOutput * (((L - L_min) / (L_max - L_min)).^a_gamma);
```

Listing 6.4. MATLAB code: The expansion method by Akyüz et al. [17].

The authors worked only with high resolution HDR images, which were artistically captured and were without compression artifacts. While this works well under such ideal conditions, in more realistic scenarios, such as television programs or conventional media where compression is employed, this may not always be the case. In these cases, a more accurate expansion needs to be performed in order to avoid amplification of compression artifacts and contouring.

Listing 6.4 provides the MATLAB code of Akyüz et al.'s EO [17]. The full code may be found in the file **AkyuzEO.m**. The method takes the following parameters as input: the maximum luminance intensity of the HDR display **maxOutput** (k in Equation (6.3)) and the nonlinear scaling factor **a_gamma** (γ in Equation (6.3)). After the preliminary steps, the maximum and the minimum luminance pixels are calculated from the variable 1 using the MATLAB functions **max.m** and **min.m**, respectively. The expansion step, which is equivalent to Equation (6.3), is then executed.

6.4.3 Gamma Expansion for Over-Exposed LDR Images

Masia et al. [251] conducted two psychophysical studies to analyze the behavior of an EO across a wide range of exposure levels. The authors then used the results of these experiments to develop an expansion technique for over-exposed content. In these experiments, three EOs were compared: Rempel et al. [324], linear expansion, and Banterle et al. [48]. The results show how the performances decrease with the increase of the exposure of the content. Furthermore, the authors observed that several artifacts, typically visible in LDR renditions of an image, were produced by EOs. These artifacts were not simply due to the incorrect intensity values but also had a spatial component. Understanding how an EO can affect the perception of the final output image may help to develop better EOs.

As shown in Figure 6.9, the authors noticed that when the increase of the exposure details are lost, this is due to the pixel saturation making the corresponding color fade to white. Based on this observation, they proposed to expand an LDR image by making the remaining details more

(a)   (b)   (c)   (d)

Figure 6.9. An example of Masia et al.'s method [251] applied to an over-exposed LDR image at different exposures: (a) Original LDR image. (b), (c), and (d) are LDR exposure images from the expanded image in (a), respectively, at f-stop -8, f-stop - 10, and f-stop -12.

prominent. This approach is different from the usual EOs approach, where saturated areas are boosted.

A straightforward way to achieve this is gamma expansion. Masia et al. proposed an automatic way to define the γ value based on the dynamic content of the input LDR image. Similarly to Akyüz and Reinhard [18], a key k value is computed as

$$k = \frac{\log L_{d,\,H} - \log L_{d,\,min}}{\log L_{d,\,max} - \log L_{d,\,max}}, \tag{6.4}$$

where $L_{d,\,H}$, $L_{d,\,min}$, and $L_{d,\,max}$ are, respectively, the logarithmic average, the minimum luminance value and the maximum luminance value of the input image. The k value is a statistic that helps clarify if the input image is subjectively dark or light. In order to predict the gamma value automatically, a pilot study was conducted where users were asked to adjust the γ value manually in a set of images. The data was empirically fitted via linear regression obtaining:

$$\gamma(k) = 10.44k - 6.282. \tag{6.5}$$

One of the major drawbacks of this expansion technique is that it may fail to utilize the dynamic range to its full extent [251]. Moreover, the fitting only works correctly on the set of test images. Some images, not belonging to the original data set, can have negative γ values; e.g., dark images. As shown in Figure 6.10(b), this may produce an unnatural appearance.

Masia et al. [252] revised their model using a multi-linear regression to improve the fitting in Equation (6.5):

$$\gamma = 2.4379 + 0.2319 \log L_{d,\,H} - 1.1228k + 0.0085p_{ov}, \tag{6.6}$$

(a)   (b)   (c)

Figure 6.10. An example where Masia et al.'s method [251] creates an incorrect $\gamma(x)$ value: (a) LDR image used in this example. (b) The image in (a) expanded ($\gamma(k) = -1.7152$) using Masia et al.'s method [251] at f-stop -12. This introduces a reciprocal that produces an unnatural appearance. (c) The new fitting [252] in this case solves the issue.

where p_{ov} is the percentage of over-exposed pixels, i.e., with values over 254. This new fitting reduces the cases when γ is negative, but it can still create them when the input image is extremely dark.

Listing 6.5 provides the MATLAB code of Masia et al.'s operator [251]. The full code can be found in the file **MasiaEO.m**. The method takes as input the LDR image, img. After the initial steps common to all EOs, the image key is calculated as in Equation (6.4) using the function imKey in the folder util. If the classic method [251] is employed, i.e., m_multi_reg == 0, each color channel will be exponentiated by m_gamma using Equation (6.5). Otherwise, i.e., m_multi_cor == 1, Equation (6.6) will be used for computing m_gamma.

```
[key, Lav]  = imKey(img);

%calculate the gamma correction value
if(m_multi_reg == 0)
    a_var = 10.44;
    b_var = -6.282;
    m_gamma = imKey(img) * a_var + b_var;
else
    %percentage of over-exposed pixels
    p_ov = length(find((L * 255) >= 254 )) / imNumPixels(L) *
        100.0;
    %Equation 5 of (2) paper
    m_gamma = 2.4379 + 0.2319 * log(Lav) - 1.1228 * key +
```

```
        0.0085 * p_ov;
end

if(m_noise)%noise removal using bilateral filter
    %note that the original paper does not provide parameters
        for filtering
    Lbase = bilateralFilter(L);
    Ldetail = RemoveSpecials(L ./ Lbase);
    Lexp = Ldetail .* (Lbase.^m_gamma);
else
    Lexp = L.^m_gamma;
end
Lexp = Lexp * maxOutput;
```

Listing 6.5. MATLAB code: The expansion methods by Masia et al. [251, 252].

Recently, Bist et al. [64] conducted a subjective study from which they found out that the lighting style of an image is strongly related to the gamma value for expansion. The result of this study led to a novel gamma formulation:

$$\gamma = 1 + \log_{10}\left(\frac{1}{L_{\mathrm{d}}^{\star}}\right), \tag{6.7}$$

where $L_{\mathrm{d}}^{\star}$ is the normalized median luminance value of the LDR image clamped in $[0.05, 0.95]$. Note that Equation (6.7) cannot suffer from negative values. The authors ran a validation experiment that showed Equation (6.7) outperformed previous gamma methods and Akyüz et al. [17] when content is display on a $1,000 \ \mathrm{cd/m^2}$ HDR television.

6.5   Classification Models

Classification methods identify or classify an image or a video frame into different parts such as diffuse regions, highlights, light source, etc. Each classified part is expanded using an appropriate method minimizing artifacts such as quantization errors, contouring, etc. Furthermore, these methods aim to avoid expanding areas that should not be expanded. For example, a light bulb should be expanded, but a white wall should not.

6.5.1   Highlights Reproduction for HDR Monitors

Meylan et al. [262, 263] presented an EO with the specific task of representing highlights in LDR images when displayed on HDR monitors. The method detects the diffuse and specular parts of the image and expands these using different linear functions. The first step of the algorithm is to calculate a threshold ω that separates highlights and diffuses regions in the luminance channel, L_{d}; see Figure 6.11. First, the image is filtered using

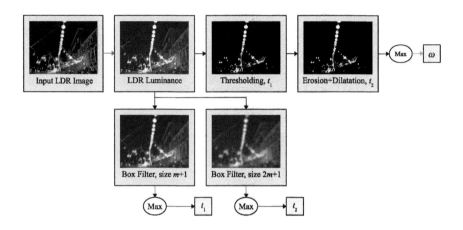

Figure 6.11. The pipeline for the calculation of the maximum diffuse luminance value ω in an image in Meylan et al. [263].

a box filter of size $m = \max(\text{width}, \text{height})/50$ to calculate value t_1 as the maximum of the filtered luminance channel. This operation is repeated for a filter of size $2m + 1$ to calculate t_2. Second, a mask, M, is computed by thresholding the original luminance with t_1. To eliminate isolated pixels, erosion and dilatation filters are applied to M using the previously computed value t_2. These can be applied over some iterations in order to obtain a stable mask. A few iterations are typically enough. At this point, pixels with the value 1 in the mask are considered specular pixels, while black pixels are considered as diffuse ones. Subsequently, ω is calculated as the minimum luminance value of the specular pixels in L_d.

After computing ω, the luminance channel is expanded using the function:

$$L_\mathrm{w}(\mathbf{x}) = f(L_\mathrm{d}(\mathbf{x})) = \begin{cases} s_1 L_\mathrm{d}(\mathbf{x}) & \text{if } L_\mathrm{d}(\mathbf{x}) \leq \omega, \\ s_1\omega + s_2(L_\mathrm{d}(\mathbf{x}) - \omega) & \text{otherwise}; \end{cases}$$

$$s_1 = \frac{\rho}{\omega} \quad s_2 = \frac{1 - \rho}{L_{\mathrm{d, max}} - \omega}, \tag{6.8}$$

where ρ is the percentage of the HDR display luminance allocated to the diffuse part that is defined by the user.

Applying f globally can lead to quantization artifacts around the enhanced highlights. These artifacts are reduced using a selective filter in the specular regions; see Figure 6.12 for the full pipeline. First, the expanded luminance, $f(L_\mathrm{d}(\mathbf{x}))$, is filtered with a 5×5 average filter obtaining f'. Second, $f(L_\mathrm{d}(\mathbf{x}))$ and f' are blended using linear interpolation and a mask,

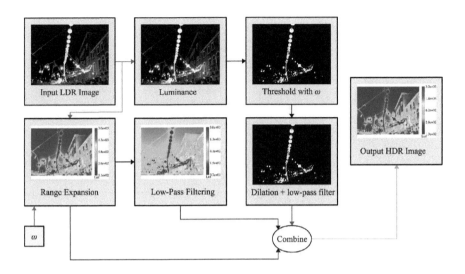

Figure 6.12. The full pipeline for the range expansion in Meylan et al.'s method [263].

which is calculated by thresholding the LDR luminance with ω and applying a dilatation and a 5×5 average filter. Finally, the authors ran a series of experiments to determine the value of ρ for f using a Brightside DR-37p HDR monitor [76]. The results showed that for outdoor scenes users preferred a high value of ρ, which means a small percentage of dynamic range allocated to highlights, while for indoor scenes the percentage was the opposite. For indoor and outdoor scenes of equal diffuse brightness, users chose a low value for ρ; so they preferred more range allocated to highlights. The analysis of the data showed that $\rho = 0.66$ is a good general estimate.

This algorithm is designed for a specific task—the reproduction of highlights on HDR monitors. The use of the method for other tasks, such as enhancement of videos, needs a classifier.

6.5.2 Enhancement of Bright Video Features for HDR Displays

Didyk et al. [113] proposed an interactive system for enhancing brightness of LDR videos, targeting and showing results for DVD content. This system was to classify a scene into three components: diffuse, reflections, and light sources, and then enhances only reflections and light sources. The authors explained that diffuse components are difficult to enhance without creating visual artifacts, and it was probably the intention of film makers to show them saturated as opposed to light sources and clipped reflections. The

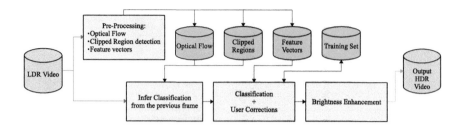

Figure 6.13. The pipeline of the system proposed by Didyk et al. [113]: prepro-cessing (calculation of features vector, optical flow, and clipped regions), classifi-cation of regions using temporal coherence and a training set, user corrections (in-cluding updating the training set), and brightness enhancement (Equation (6.9)).

system works on nonlinear values because the goal is enhancement and not physical accuracy.

The system consists of three main parts: preprocessing, classification, and enhancement of clipped regions. The pipeline can be seen in Fig-ure 6.13. The preprocessing step generates data needed during the clas-sification. In particular, it determines clipped regions using a flood-fill algorithm. At least one channel must be saturated (over 230 for DVD content), and luma values must be greater than 222. Also, at this stage, optical flow is calculated as well as other features such as image statistics, geometric features, and neighborhood characteristics.

Classification determines lights, reflections, and diffuse regions in a frame and relies on a training set of 2,000 manually classified regions. Pri-marily, a support vector machine [389] with kernel $k(z, z') = \exp(-\gamma \|z - z'\|^2)$ performs an initial classification of the regions. Subsequently, motion tracking improves the initial estimation, using a nearest neighbor classifier based on an Euclidean metric:

$$d^2((z, x, t), (z', x', t')) = 50\|z - z'\|^2 + \|z - z'\|^2 + 5(t - t')^2,$$

where z are region features, x are coordinates in the image, and t is the frame number. This is allowed to reach a classification error of 12.6% on all regions used in the tests. Tracking of clipped regions using motion com-pensation further reduced the percentage of objects that require manual correction to 3%. Finally, the user can supervise the classified regions, correcting wrong classifications using an intuitive user interface; see Fig-ure 6.14.

Clipped regions are enhanced by applying a nonlinear adaptive tone curve. This is calculated based on partial derivatives within a clipped region stored in a histogram H. The tone curve is defined as a histogram

Figure 6.14. The interface used for adjusting classification results in Didyk et al.'s system [113]. (The image is courtesy of Piotr Didyk.)

equalization on the inverted values of H as

$$f(b) = k \sum_{j=2}^{b} (1 - H[j]) + t_2, \tag{6.9}$$

where t_2 is the lowest luma value for a clipped region, and k is a scale factor that limits to the maximum boosting value m (equal to 150% for lights and 125% for reflections):

$$k = \frac{m - t_2}{\sum_{j=1}^{N} (1 - H[j])},$$

where N is the number of bins in H. To avoid contouring during boosting, the luma channel is filtered with bilateral filtering, separating it into fine details and a base layer, which is merged after luma expansion. The method is semi-automatic because intervention of the user is required.

6.6 Expand Map Models

In over-exposed regions of an image/frame, information has been lost, and it can not be reconstructed anymore. Researchers have made attempts to approximate this lost information by interpolating/multiplying an expanded LDR image/frame by a guidance map, which approximate the lost low frequency profile of luminance in over-exposed areas. Following the terminology used in Banterle et al. [48], these guidance methods are referred to as *expand maps*.

6.6.1 Density Estimation based Expand Maps

A general framework for expanding LDR content for HDR monitors and IBL was proposed by Banterle et al. [48, 50]. The key points of the framework are the use of an inverted TMO for expanding the range combined with a smooth field for the reconstruction of the lost over-exposed areas.

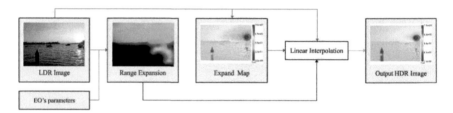

Figure 6.15. The pipeline of Banterle et al.'s method [48, 50].

The first step of the framework is to linearize the input image. Figure 6.15 shows the pipeline. If the CRF is known, its inverse is applied to the signal. Otherwise, single image methods can be employed; see Section 6.2.2. Subsequently, the range of the image is expanded by inverting a TMO. In Banterle et al.'s implementation, the inverse of the global Reinhard et al.'s operator [323] was used. This is because the operator has only two parameters, and range expansion can be controlled in a straightforward way. This inverted TMO is defined as

$$L_{w'}(\mathbf{x}) = \frac{1}{2} L_{w,\,max} L_{white} \left(L_d(\mathbf{x}) - 1 + \sqrt{\left(1 - L_d(\mathbf{x})\right)^2 + \frac{4}{L_{white}^2} L_d(\mathbf{x})} \right),$$

(6.10)

where $L_{w,\,max}$ is the maximum output luminance in cd/m^2 of the expanded image, and $L_{white} \in (1, +\infty)$ is a parameter that determines the shape of the expansion curve. This is proportional to the contrast. The authors suggested a value of $L_{white} \approx L_{w,\,max}$ to increase the contrast while limiting artifacts due to expansion.

After range expansion, the expand map, E, is computed. The expand map is a smooth field representing a low frequency version of the image in areas of high luminance. It has two main goals. The first one is to reconstruct lost luminance profiles in over-exposed areas of the image. The second goal is to attenuate quantization or compression artifacts that can be enhanced during expansion.

(a)   (b)

Figure 6.16. Application of Banterle et al.'s method [48, 50] for relighting synthetic objects: (a) Lucy's model is relit using the St. Peter's HDR environment map. (b) Lucy's model is relit using an expanded St. Peter's LDR environment map (starting at exposure 0). Note that colors in (a) and (b) are close; this means that they are well reconstructed. Moreover, reconstructed shadows in (b) follow the directions of the ones in (a), but they are less soft and present some aliasing. (The original St. Peter's HDR environment map is courtesy of Paul Debevec. The Lucy's model is courtesy of the Stanford 3D models repository.)

The expand map is implemented by applying density estimation on samples, $\mathcal{S}$, (with an intensity and a position) generated using importance sampling (e.g., median-cut sampling [108]):

$$E(\mathbf{x}) = \sum_{p \in \mathcal{P}(\mathbf{x})} \Psi_p \gamma \left(1 - \frac{1 - e^{-\beta \frac{d_p^2}{2r_{\max}^2}}}{1 - e^{-\beta}} \right), \qquad (6.11)$$

where $\mathcal{P}$ is the set of samples in $\mathcal{S}$ within $r_{\max}$ from $\mathbf{x}$, Ψ_p is the intensity of the p-th sample, d_p is the distance from $\mathbf{x}$ to the p-th sample, and $\gamma = 0.918$ and $\beta = 1.953$ [48]. Finally, the expanded LDR image and the original one are combined as

$$L_{\mathrm{w}}(\mathbf{x}) = L_{\mathrm{w}'}(\mathbf{x}) E(\mathbf{x}) + L_{\mathrm{d}}(\mathbf{x}) \big(1 - E(\mathbf{x}) \big). \qquad (6.12)$$

Equation (6.12) keeps low luminance values as in the original value, which avoids compression for low values when L_{white} is set to a high value (around 10^4 or more). Otherwise, artifacts such as contours could appear.

Banterle et al. [49] extended their framework for automatically processing images and videos by using a three-dimensional sampling algorithms,
volume density estimation, and heuristics for determining the parameters.
Furthermore, they showed that cross bilateral filtering the expand map by
the input LDR luminance reduces distortions at edges. This framework can
run in real-time, more than 24 fps on HD content, by using point-based
graphics on modern GPUs [36, 43].

This algorithm presents a general solution for visualization on HDR
monitors and IBL; see Figure 6.16. Furthermore, the authors tested it using
HDR-VDP [243] for both tasks to prove its efficiency compared with the
more straightforward exposure methods. The main limit of the framework
is that large over-exposed areas, i.e., more than 30% of the image, can
not be reconstructed using an expand map because there is not enough
information to exploit.

6.6.2   Edge Stopping Expand Map

Rempel et al. [324] proposed a similar technique based on expand maps.
Their goal was real-time LDR expansion for videos and images. Figure 6.17
shows the the full pipeline of the algorithm.

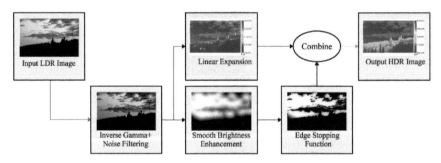

Figure 6.17. The pipeline of Rempel et al.'s method [324].

The first step of the algorithm is to remove artifacts due to the compression algorithms by using a bilateral filter with small intensity and spatial
kernels. Sophisticated artifacts removal is not employed due to real-time
constraints. The next step of the method is to linearize the signal using an
inverse gamma function. Then, the contrast is linearly expanded as

$$L_w(\mathbf{x}) = L_d(\mathbf{x})(L_{w,\,max} - L_{w,\,min}) + L_{w,\,max},\qquad(6.13)$$

where $L_{w,\,max} = 1200$ cd/m^2 and $L_{w,\,min} = 0.3$ cd/m^2 are optimized values
for the Brightside DR-37p HDR monitor [76] in order to have a maximum

Figure 6.18. An example of Rempel et al.'s method [324]: (a) The input LDR image. (b), (c), (d) Different f-stops after expansion.

contrast up to about 5,000:1 to avoid artifacts. To enhance brightness in bright regions, a brightness enhance function (BEF) is employed, i.e., an expand map. This function is calculated by applying a threshold of 0.92 (on a scale $[0, 1]$ for LDR values). At this point, the image is filtered using a Gaussian filter with a $\sigma = 30$ (i.e., 150 pixels), which is meant for 1920×1080 content. In order to increase contrast around edges, an edge stopping function is used. Starting from saturated pixels, a flood-fill algorithm strategy is applied until an edge is reached; this is estimated using gradients. Subsequently, a morphological operator followed by a Gaussian filter with a smaller kernel is applied to remove noise. The BEF is mapped in $[1, \alpha]$ where $\alpha = 4$ and is finally multiplied with the scaled image, Equation (6.13), to generate the final HDR image; see Figure 6.18.

To improve efficiency, the BEF is calculated using Laplacian pyramids [81], which can be implemented on a GPU or FPGA efficiently. The algorithm was evaluated using HDR-VDP [243] comparing the linearized starting image with the generated HDR image. This evaluation was needed to show that the proposed method does not introduce spatial artifacts during expansion of the content. Note that this method processes each frame separately and may not be temporally coherent (even though parameters are fixed for all frames) due to the nature of the edge stopping function.

6.6.3   Bilateral Filter–Based Solutions

Kovaleski and Oliveira [201,202] proposed a method that expands the content as in Rempel et al.'s method [324], but the expand map is computed as the result of the application of a cross bilateral filter between a binary map indicating over-exposed pixels, T, and the input LDR luminance:

$$E(\mathbf{x}) = CBF[T, L_{\mathrm{d}}](\mathbf{x}, f_s, g_r), \qquad (6.14)$$

where CBF is a cross bilateral filter (see Appendix A), f_s is a Gaussian function with $\sigma_s = 150$ (for HD content), g_r is a Gaussian function with $\sigma_r = 25/255$, and T is defined as

$$T(\mathbf{x}) = \begin{cases} 1 & \text{if } A(\mathbf{x}) > t, \\ 0 & \text{otherwise;} \end{cases} \quad A(\mathbf{x}) = \max\big(I_R(\mathbf{x}), I_G(\mathbf{x}), I_B(\mathbf{x})\big),$$

where I is the input image (subscripts R, G, and B are, respectively the red, green, and blue color channels), and t is a threshold depending on the content ($t = 230/255$ for images, and $t = 254/255$ for videos). To speed up computations in Equation (6.15), the method uses the bilateral grid data structure [89] on the GPU. This achieves real-time performance, around 25 fps, on full HD content and fewer distortions than Rempel et al.'s method [324].

```
imgC = zeros(size(img,1), size(img, 2));
imgC(max(img, [], size(img,3)) > threshold) = 1;

e_map = bilateralFilter(imgC, L, 0, 1, ko_sigma_s, ko_sigma_r);
e_map = e_map * 3.0 + 1.0; %scale it in [1, 4]

%compute expanded luminance
Lexp = L * (ko_display_max - ko_display_min) + ko_display_min;
Lexp = Lexp .* e_map; %scale it by e_map
```

Listing 6.6. MATLAB code: The code of Kovaleski and Olivera's EO [202].

Listing 6.6 provides the MATLAB code of the Kovaleski et al.'s operator [202]. The full code can be found in the file KovaleskiOliveira.m. The method takes as input the LDR image, img, parameters for the linear expansion ko_display_max (minimum output luminance) and ko_display_max (maximum output luminance), parameters for the bilateral grid (ko_sigma_s and ko_sigma_r), and type_content that determines the threshold for classifying over-exposed pixels depending on the nature of the content. After the initial steps common to all EOs, it creates T, imgC, that is filtered using a cross bilateral grid bilateralFilter.m (in the util folder) obtaining the expand map, e_map. Finally, the input LDR image is linearly expanded, Lexp, and multiplied by e_map.

Huo et al. [179] introduced an EO where the expand map, E, is the application of the bilateral filter on the whole luminance channel without thresholding as in Kovaleski and Oliveira:

$$E(\mathbf{x}) = BF[L_{\mathrm{d}}^1](\mathbf{x}, f_{s2}, g_{r2}) \quad L_{\mathrm{d}}^1(\mathbf{x}) = BF[L_{\mathrm{d}}](\mathbf{x}, f_{s1}, g_{r1}), \qquad (6.15)$$

where BF is a bilateral filter (see Appendix A), and f_{s1}, g_{r1}, f_{s1}, and g_{r2} are Gaussian functions with, respectively, $\sigma_{s1} = 16$, $\sigma_{r1} = 0.3$, $\sigma_{s1} = 10$, $\sigma_{r1} = 0.1$.

Finally, the expanded luminance is computed inverting the sigmoid of the retina response to have a physiologically based operator, Equation (1.1):

$$L_{\mathrm{w}}(\mathbf{x}) = \left(L_{\mathrm{d}}(\mathbf{x})\left(E(\mathbf{x})L_{\mathrm{w,\,max}} + (L_{\mathrm{d,\,H}}L_{\mathrm{w,\,max}})^n\right)\right)^{\frac{1}{n}} \qquad (6.16)$$

where $n = 0.83$. In their evaluation based on the dynamic range independent quality assessment metric (see Section 10.1.3), the authors showed that their method can expand LDR content with negligible contrast loss and contrast reversal when compared to the state-of-the-art at that time (i.e., linear expansion [17,263], gamma expansion [48,251], and expand map methods [324]).

```
%compute image statistics
sigma_1 = logMean(L);

%iterative bilateral filter: 2 passes as in the original paper
L_s_1_1 = bilateralFilter(L, [], 0.0, 1.0, 16.0, 0.3);
L_s_1_2 = bilateralFilter(L_s_1_1, [], 0.0, 1.0, 10.0, 0.1);

%compute parameters
sigma = maxOutput * sigma_1;
L_s_h = maxOutput * L_s_1_2;

%expand luminance
Lexp = ((L / max(L(:))) .* ((L_s_h.^hou_n + sigma^hou_n)).^(1.0
    / hou_n));
```

Listing 6.7. MATLAB code: The code of Huo et al.'s EO [179].

Listing 6.7 provides the MATLAB code of the Huo et al.'s operator [179]. The full code can be found in the file HuoPhysEO.m. The method takes as input the LDR image, img, maximum output luminance, maxOutput, and the sigmoid exponent hou_n. After the initial steps common to all EOs, the method computes the geometric mean of luminance,

sigma_1. Then, luminance is filtered two times using a bilateral grid, **bilateralFilter.m** (in the **util** folder), obtaining the expand map, **L_s_1_2**. Both **L_s_1_2** and **sigma_1** are scaled by **maxOutput**, and Equation (6.16) is evaluated obtaining **Lexp**.

6.7 User-Based Models: HDR Hallucination

Since it may not always be possible to recover missing HDR content using automatic approaches, Wang et al. [402] proposed a system where users can add details to the expanded LDR content. They termed the whole process *hallucination*, and their system presents a mixture between an automatic and a user-based approach.

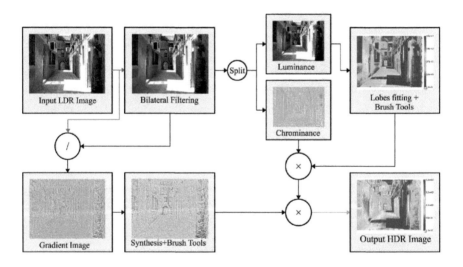

Figure 6.19. The pipeline of the Wang et al. [402] method.

The first step of hallucinating an LDR image is to linearize its signal; see Figure 6.19 for the full pipeline of the system. Linearization involves the application of an inverse gamma function with $\gamma = 2.2$; this is the standard value for DVDs and television formats [183]. After this step, the image is decomposed into large-scale illumination and fine texture details. A bilateral filter is applied to the image I obtaining a filtered version I_f. The texture details are obtained as $I_d = I/I_f$. Radiance for large-scale illumination I_f is estimated via elliptical Gaussian fitting. To achieve this, a weight map, w, is calculated for each pixel as

$$
w(\mathbf{x}) = \begin{cases} \frac{C_{ue}-Y(\mathbf{x})}{C_{ue}} & Y(\mathbf{x}) \in [0, C_{ue}), \\ 0 & Y(\mathbf{x}) \in [C_{ue}, C_{oe}), \\ \frac{Y(\mathbf{x})-C_{oe}}{1-C_{oe}} & Y(\mathbf{x}) \in [C_{oe}, 1], \end{cases}
$$

where $Y(\mathbf{x}) = R_s(\mathbf{x}) + 2G_s(\mathbf{x}) + B_s(\mathbf{x})$, and C_{ue} and C_{oe} are, respectively, the thresholds for under-exposed and over-exposed pixels. The authors suggested values of 0.05 and 0.85 for C_{ue} and C_{oe}, respectively. Then, each over-exposed region is segmented and fitted with an elliptical Gaussian lobe G, where variance of the axis is estimated using region extents, and the profile is calculated using an optimization procedure based on non-over-exposed pixels at the edge of the region. The luminance is blended using a simple linear interpolation:

$$
O(\mathbf{x}) = w(\mathbf{x})G(\mathbf{x}) + (1 - w(\mathbf{x}))\log_{10} Y(\mathbf{x}).
$$

Optionally, users can add Gaussian lobes using a brush. The texture details, I_d, are reconstructed using a texture synthesis technique similar to Bertalmio et al. [59], where the user can select an area as a source region by drawing it with a brush. This automatic synthesis has some limits when scene understanding is needed; therefore, a warping tool is included. This allows the user to select, with a stroke-based interface, a source region and a target region; pixels will be transferred. This is a tool similar to the stamp and healing tools in Adobe Photoshop [7]. Finally, the HDR image is built by blending the detail and the large-scale illumination using Poisson image editing [305] in order to avoid seams in the transition between expanded over-exposed areas and well-exposed areas.

This system can be used for both IBL and visualization of images, and compared with other algorithms it may maintain details in clamped regions. However, the main problem of this approach is that it is user-based and not automatic, which potentially limits its use to single images and not videos. Recently, Kuo et al. [210] used hierarchical inpainting to make the system fully automatic, but they did not provide temporal coherent mechanisms for hallucinating videos.

6.8   Summary

An overview of all methods discussed is presented in Table 6.1. This summarizes what techniques are used, and their applicability to videos. Most of the methods expand the dynamic range using either a linear, or non-linear function, while Meylan et al. use a two-scale linear function. The reconstruction methods aim to smoothly expanding the dynamic range and

Method	Expansion Function	Reconstruction + Noise Reduction	Temporal stable
Landis [213]	Non-Linear	N/A	no
Akyüz et al. [17]	Linear	N/A	yes[+]
Masia et al. [251]	Non-Linear	N/A	yes[+]
Bist et al. [64]	Non-Linear	N/A	yes[+]
Meylan et al. [263]	Linear	Bilateral Filter	no
Didyk et al. [113]	Non-Linear	Bilateral Filter	yes
Banterle et al. [43]	Non-Linear	Expand Map	yes
Rempel et al. [324]	Linear	Expand Map	yes
Kovaleski and Oliveira [202]	Linear	Expand Map	yes
Huo et al. [179]	Non-Linear	Expand Map	yes[+]
Wang et al. [402]	Non-Linear	Inpainting	no

Table 6.1. Classification of algorithms for expansion of LDR content for HDR applications. [+] means that the method may be temporally coherent by computing statistics temporally.

a variety of methods are proposed. Section 10.4 provides discussion on evaluation methods for expansion operators but some of the findings are summarized here for completeness.

Recently, expansion operators have increased their popularity thanks to the introduction of HDR displays. This is because there is the need to correctly process and visualize the huge amount of legacy content within the HDR pipeline. There still no consensus which is the best operator for which task. However, studies on EOs highlighted important facts:

- Akyüz et al. [17] showed that a global method can provide an HDR experience to users when content is well exposed and not compressed.

- Banterle et al. [47] pointed out that non-linearity is important, especially for IBL, and mechanism to reduce quantization and contouring artifacts are extremely important for still images and videos when they have been compressed.

- Masia et al. [251] showed that when the number of over-exposed pixels increases (i.e., very large areas), users prefer global expansion methods because reconstruction ones (i.e., expand maps-based) can introduce visible artifacts.

- De Simone et al. [104] noted that many EOs are not meant for videos, and they introduce visible temporal artifacts such as global and local flickering. Therefore, temporal coherency is an important feature when dealing with videos.

- Abebe et al. [1] highlighted that there is the need of color management in an EO and that this is missing in most of them.

These studies also provide future research directions, and most importantly new research questions may arise such as *How can we achieve high quality and artifact-free LDR videos expansion?* and *How should we handle colors during expansion?*

7
Image-Based Lighting

Since HDR images may represent the true physical properties of lighting at a given point, they can improve the rendering process. In particular, a series of techniques commonly referred to as *image-based lighting* (IBL) provide methods of accelerating the rendering computations lit by HDR images. These HDR images are, almost always, obtained by capturing the entire sphere of lighting at a point in a real scene. IBL methods render content by using the captured HDR image as lighting in shading computations, recreating the same physical lighting conditions in the virtual scene as in the real scene. This results in realistic-looking images that have been embraced by the film and games industries.

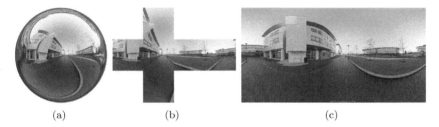

(a)	(b)	(c)

Figure 7.1. The Computer Science environment map encoded using different projection mappings: (a) Angular map. (b) Cube-map unfolded into a horizontal cross. (c) Latitude-longitude map.

7.1 Environment Maps

IBL usually takes as input an image, termed the environment map, that has captured illumination of the real-world environment for each direction, $\mathbf{D} = [x, y, z]^\top$, around a point. Therefore, an environment map can be

parameterized on a sphere. Different two-dimensional projection mappings
of a sphere can be adopted to encode the environment map. The most
popular methods used in computer graphics are: angular mapping (Figure 7.1(a)), cube mapping (Figure 7.1(b)), and latitude-longitude mapping
(Figure 7.1(c)).

Listing 7.1 provides the main MATLAB code for converting between
one projection and another.

```
function imgOut = ChangeMapping(img, mappingIn, mappingOut)
...
%first step: generate directions
[r, c, col] = size(img);

if(strcmpi(mappingIn, mappingOut) == 0)
    maxCoord = max([r, c]) / 2;
    switch mappingOut
        case 'LongitudeLatitude'
            D = LL2Direction(maxCoord, maxCoord * 2);
        case 'Angular'
            D = Angular2Direction(maxCoord, maxCoord);
        case 'CubeMap'
            D = CubeMap2Direction(maxCoord * 4, maxCoord * 3);
        otherwise
            error([mappingOut, ' is not valid.']);
    end

    %second step: values' interpolation
    switch mappingIn
        case 'LongitudeLatitude'
            [X1,Y1] = Direction2LL(D, r, c);
        case 'Angular'
            [X1,Y1] = Direction2Angular(D, r, c);
        case 'CubeMap'
            [X1,Y1] = Direction2CubeMap(D, r, c);
        otherwise
            error([mappingIn, ' is not valid.']);
    end
    X1 = real(round(X1));
    Y1 = real(round(Y1));

    %interpolation
    [X, Y] = meshgrid(1:c, 1:r);
    imgOut = [];
    for i=1:col
        imgOut(:,:,i) = interp2(X, Y, img(:,:,i), X1, Y1, '*
            cubic');
    end

    [rM, cM, ~] = size(imgOut);
    switch mappingOut
        case 'CubeMap'
            imgOut = imgOut .* CrossMask(rM, cM);
```

```
        case 'Angular'
            imgOut = imgOut .* AngularMask(rM, cM);

        case 'Spherical'
            imgOut = imgOut .* AngularMask(rM, cM);
    end
    imgOut = RemoveSpecials(imgOut);
else %if it is the same mapping no work to be done
    imgOut = img;
end
end
```

Listing 7.1. MATLAB code: The main code for changing an environment map from one two-dimensional projection to another.

The function ChangeMapping.m, which can be found under the EnvironmentMaps folder, takes as input the environment map to be converted img. Two strings representing the original mapping and the one to be converted to, via the parameters mapping1 and mapping2, respectively, are also passed as parameters. This function operates by first converting the original mapping into a series of directions via the functions LL2Direction.m, Angular2Direction.m, and CubeMap2Direction.m, which can be found under the EnvironmentMaps folder. These represent conversions from original maps stored using latitude-longitude, angular, and cube-map representations, respectively. The second step converts from the sets of directions toward the second mapping using the functions Direction2LL.m, Direction2Angular.m, and Direction2CubeMap.m, which can be found under the EnvironmentMaps folder. Finally, bilinear interpolation is used for the final images. Note that some mapping methods, such as angular and cube mapping, generate images that do not cover the full area of a rectangular image. During interpolation, the empty areas are set to invalid values (i.e., NaN or Inf float values). In order to remove these invalid values, some masks are generated. These masks are created using the functions CrossMask.m and AngularMask.m that are, respectively, used for the cube and angular mapping methods. These functions may be found under the EnvironmentMaps folder.

7.1.1   Latitude-Longitude Mapping

A typical environment mapping is the so-called latitude-longitude mapping, which maps the sphere into a rectangular domain. The mapping from rectangular domain into directions on the sphere is given as:

$$
\begin{bmatrix} D_x \\ D_y \\ D_z \end{bmatrix} = \begin{bmatrix} \sin\phi\sin\theta \\ \cos\theta \\ -\cos\phi\sin\theta \end{bmatrix}, \qquad \begin{bmatrix} \phi \\ \theta \end{bmatrix} = \begin{bmatrix} 2\pi x \\ \pi(y-1) \end{bmatrix}, \tag{7.1}
$$

where $x \in [0,1]$ and $y \in [0,1]$. Equation (7.1) transforms texture coordinates $[x, y]^T$ to spherical ones $[\phi, \theta]^T$ and finally to direction coordinates $[D_x, D_y, D_z]^T$. Listing 7.2 provides the MATLAB code for converting from a latitude-longitude representation to a set of directions following the above equation.

```
function D = LL2Direction(r, c)

[X0, Y0] = meshgrid(0:(c - 1), 0:(r - 1));
phi   =   pi * ((X0 / c) * 2 - 1) - pi * 0.5;
theta =   pi * (Y0 / r);

sinTheta = sin(theta);
D = zeros(r, c, 3);
D(:,:,1) =   cos(phi)   .* sinTheta;
D(:,:,2) =   cos(theta);
D(:,:,3) =   sin(phi)   .* sinTheta;

end
```

Listing 7.2. MATLAB code: Converting from a latitude-longitude representation to a set of directions as in Equation (7.1).

The main advantage of this mapping is that it is easy to understand and implement. However, it is not an equal-area mapping since pixels cover different areas on the sphere. For example, pixels at the equators cover more area than pixels at the poles. This problem has to be taken into account when these environment maps are sampled.

The inverse mapping, from the direction on the sphere into the rectangular domain, is given as:

$$\begin{bmatrix} x \\ y \end{bmatrix} = \begin{bmatrix} 1 + \frac{1}{\pi} \arctan(D_x, -D_z) \\ \frac{1}{\pi} \arccos D_y \end{bmatrix}. \tag{7.2}$$

Listing 7.3 provides the MATLAB code for converting the set of directions into a latitude-longitude representation. The initial image resize in the listing adjusts the input coordinates to have the same aspect ratio as the output mapping. The rest of the code implements Equation (7.2).

```
function [X1, Y1] = Direction2LL(D, r, c)
X1 = 1 + atan2(D(:,:,1), -D(:,:,3)) / pi;
Y1 = acos(D(:,:,2)) / pi;
X1 = RemoveSpecials(X1 * c / 2);
Y1 = RemoveSpecials(Y1 * r);
end
```

Listing 7.3. MATLAB code: Converting from a set of directions to a latitude-longitude representation as in Equation (7.2).

7.1.2 Angular Mapping

Another type of environment mapping is angular mapping. This maps a sphere into a two-dimensional circular area. The mapping from the two-dimensional circular area into directions on the sphere is:

$$
\begin{bmatrix} D_x \\ D_y \\ D_z \end{bmatrix} = \begin{bmatrix} \cos\theta\sin\phi \\ \sin\theta\sin\phi \\ -\cos\phi \end{bmatrix}, \qquad \begin{bmatrix} \theta \\ \phi \end{bmatrix} = \begin{bmatrix} \arctan(1-2y, 2x-1) \\ \pi\sqrt{(2x-1)^2+(2y-1)^2} \end{bmatrix}, \qquad (7.3)
$$

where $x \in [0,1]$ and $y \in [0,1]$. The inverse mapping is:

$$
\begin{bmatrix} x \\ y \end{bmatrix} = \frac{1}{2} + \arccos\left(\frac{-D_z}{\pi 2\sqrt{D_x^2 + D_y^2}} \right) \begin{bmatrix} D_x \\ D_y \end{bmatrix}. \qquad (7.4)
$$

This mapping avoids under-sampling at the edges of the circle, but it is not an equal-area mapping, as in the case of latitude-longitude mapping.

The functions `Angular2Direction.m` and `Direction2Angular.m` implement Equation (7.3) and Equation (7.4), respectively. These functions are not shown in the book and we refer the reader to the implementation of the functions in the HDR Toolbox for further details.

7.1.3 Cube Mapping

Another popular mapping is cube mapping, which maps a sphere onto a cube. The cube is usually represented as six different images or an open cube on a two-dimensional plane shaped as a cross. In this book, we use the cross-shape representation. Since equations for each face are similar but with only small changes, we present equations for a single face (the front face of the cube). The other equations can be computed by swizzling coordinates and signs from the presented equations. The mapping from the front face into the directions on the sphere is:

$$
\begin{bmatrix} D_x \\ D_y \\ D_z \end{bmatrix} = \frac{1}{\sqrt{1+(2x-1)^2+(2y-1)^2}} \begin{bmatrix} 2x-1 \\ 2y-1 \\ 1 \end{bmatrix}, \text{ if } x \in \left[\frac{1}{3}, \frac{2}{3}\right] \wedge y \in \left[\frac{1}{2}, \frac{3}{4}\right].
$$

The inverse mapping is:

$$
\begin{bmatrix} x \\ y \end{bmatrix} = \frac{3}{2} \begin{bmatrix} \frac{-D_x}{2D_z} \\ \frac{D_y}{2D_z} \end{bmatrix}, \qquad \text{if } (D_z < 0) \wedge (D_z \le -|D_x|) \wedge (D_z \le -|D_y|).
$$

The main advantage of cube mapping is that it is straightforward to implement on graphics hardware. Only integer and Boolean operations are needed to fetch pixels. The main disadvantage is that it is not an equal-area mapping. Therefore, sampled pixels need to be weighted before they can be used.

Details of `CubeMap2Direction.m` and `Direction2CubeMap.m`, based on the above equations, are not given in the book but are part of the HDR Toolbox. Please refer to this for details.

7.2   Rendering with IBL

IBL was used to simulate perfect specular effects such as pure specular reflection and refraction in the seminal work by Blinn and Newell [66].

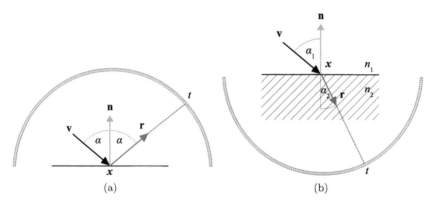

(a)   (b)

Figure 7.2. The basic Blinn and Newell [66] method for IBL: (a) The reflective case: the view vector $\mathbf{v}$ is reflected around normal $\mathbf{n}$ obtaining vector $\mathbf{r} = \mathbf{v} - 2(\mathbf{n} \cdot \mathbf{v})\mathbf{n}$, which is used as a look-up into the environment map to obtain the color value t. (b) The refractive case: the view vector $\mathbf{v}$ coming from a medium with index of refraction n_1 enters in a medium with index of refraction $n_2 < n_1$. Therefore, $\mathbf{v}$ is refracted following Snell's law, $n_1 \sin \alpha_1 = n_2 \sin \alpha_2$, obtaining $\mathbf{r}$. This vector is used as a look-up into the environment map to obtain the color value t.

It must be noted that at the time of that publication, the reflections were limited to LDR images; however, the method applies directly to HDR images. The reflected/refracted vector at a surface point $\mathbf{x}$ is used as a look-up into the environment map, and the color at that address is used as the reflected/refracted value; see Figure 7.2. This method allows very fast reflection/refraction; see Figure 7.3(a) and Figure 7.3(b). However, there are some drawbacks. First, concave objects cannot have internal interreflections/refractions, because the environment map does not take into account local features; see Figure 7.4(a). Second, reflection/refraction can be distorted since there is a parallax between the evaluation point and the point where the environment map was captured; see Figure 7.4(b).

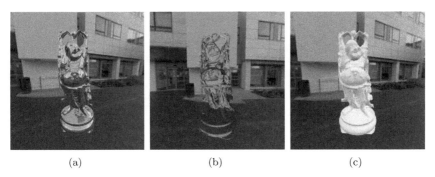

(a)   (b)   (c)

Figure 7.3.   An example of classic IBL using environment maps applied to Stanford's Happy Buddha model [154]: (a) Simulation of a reflective material. (b) Simulation of a refractive material. (c) Simulation of a diffuse material.

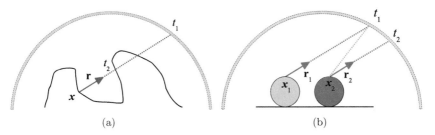

(a)   (b)

Figure 7.4.   The basic Blinn and Newell [66] method for IBL: (a) The point $\mathbf{x}$ inside the concavity erroneously uses t_1 instead of t_2 as color for refraction/reflection. This is due to the fact that the environment map does not capture local features. (b) In this case, both reflected/refracted rays for the blue and red objects are pointing in the same direction but from different starting points. However, the evaluation does not take into account the parallax, so $\mathbf{x}_1$ and $\mathbf{x}_2$ share the same color t_1.

In parallel, Miller and Hoffman [269] and Green [151] extended IBL for simulating diffuse effects; see Figure 7.3(c). This was achieved by convolving the environment map with a low-pass kernel:

$$E(\mathbf{n}) = \int_{\Omega(\mathbf{n})} L(\boldsymbol{\omega})(\mathbf{n} \cdot \boldsymbol{\omega}) d\boldsymbol{\omega}, \tag{7.5}$$

where L is the environment map, $\mathbf{n}$ is a direction in the environment map, and $\Omega(\mathbf{n})$ is the positive hemisphere of $\mathbf{n}$. An example of a convolved environment map is shown in Figure 7.5. In this case, the look-up vector

(a)   (b)

Figure 7.5. The Computer Science environment map filtered for simulating diffuse reflections: (a) The original environmental map. (b) The convolved environment map using Equation (7.5).

for a point is the normal. Nevertheless, this extension inherits the same problems of Blinn and Newell's method, namely, parallax issues and no interreflections.

Debevec [107] proposed a general method for IBL that takes into account arbitrary BRDFs and interreflections. In addition, environment maps composed solely of HDR images that, as already noted, encode real-world irradiance data were used. The proposed method is based on ray tracing; see Section 2.1.4. The evaluation, for each ray shot through a given pixel, is divided into the following cases:

1. No intersections. The ray does not hit an object in its traversal of the scene. In this case, the color of the pixel is set to the one of the environment map using the direction of the ray as a look-up vector.

2. Pure specular. The ray intersects an object with a pure specular material, such that, the ray is reflected and/or refracted according to the material properties.

3. General material. The ray intersects an object with a general material described by a BRDF. For such an occurance, a modification of the Rendering Equation [186] is evaluated as

$$L(\mathbf{x}, \boldsymbol{\omega}) = L_e + \int_\Omega L(\boldsymbol{\omega}') f_r(\boldsymbol{\omega}', \boldsymbol{\omega}) V(\mathbf{x}, \boldsymbol{\omega}') \mathbf{n} \cdot \boldsymbol{\omega}' d\boldsymbol{\omega}', \qquad (7.6)$$

where $\mathbf{x}$ and $\mathbf{n}$ are, respectively, the position and normal of the hit object; L_e is the emitted radiance at point $\mathbf{x}$, L is the environment map, f_r is the BRDF, $\boldsymbol{\omega}'$ is the out-going direction, and $\boldsymbol{\omega}$ is the view vector. The function V is the visibility function; a Boolean function that determines if a ray is obstructed by an object or not.

The use of ray tracing removes the interreflections/refractions limitations of the first approaches. Furthermore, visibility is evaluated allowing for shadows and a generally more realistic visualization; see Figure 7.6.

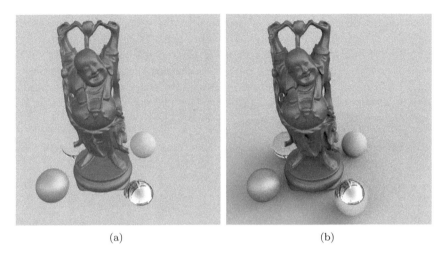

(a)   (b)

Figure 7.6. An example of IBL evaluating visibility, applied to Stanford's Happy Buddha Model [154]: (a) IBL evaluation without shadowing. (b) IBL evaluation with Debevec's method [107].

The evaluation of IBL using Equation (7.6) is computationally very expensive. Evaluation requires selecting a significant number of directions over the environment map; however a large number of these may be providing a minor contribution which would result in poor use of resources. To lower the computational complexity, two general methods, that carefully select were to best sample the environment map, are employed: light source generation or Monte-Carlo integration.

7.2.1 Light Source Generation

A commonly used set of methods for the evaluation of Equation (7.6) is to generate a finite set of directional light sources from the environment map. The fundamental concept behind these methods is the placing of a light source in the location of areas on the environment map corresponding with high luminance values. When rendering, these light sources are directly used for the illumination.

A number of techniques based on light source generation have been proposed. The most popular are: structured importance sampling (SIS) [8], k-means sampling (KMS) [198], Penrose tiling sampling (PTS) [293], and median cut sampling (MCS) [108]. The main difference in all these meth-

ods lies in where the generated light sources are placed. In SIS, a light is
placed in the center of a stratum generated by a k-center on a segmented
environment map. In KMS, lights are generated randomly on the environment map and then they are relaxed using Lloyds's method [232]. In PTS,
the image is decomposed using Penrose tiles, where smaller tiles are applied
to areas with high luminance, and a light source is placed for each vertex
of a tile. Finally, in MCS, the image is hierarchically decomposed into a
two-dimensional tree that recursively subdivides the area into regions of
equal luminance (or equal variance [393]) until there are as many regions
as the number of light sources required. The light sources are placed in the
weighted center of each region.

```
function [imgOut, lights] = MedianCut(img, nlights, falloff)
...
global L;
global imgWork;
global lights;

%falloff compensation
if(falloff)
    img = FallOffEnvMap(img);
end

L = lum(img);
[r, c] = size(L);

if(nlights < 0)
    nlights = 2.^(round(log2(min([r, c])) + 2));
end

%global variables initialization
imgWork = img;
lights = [];
MedianCutAux(1, c, 1, r, round(log2(nlights)));
imgOut = GenerateLightMap(lights, c, r);

end
```

Listing 7.4. MATLAB code: Median cut for light source generation.

Listing 7.4 shows MATLAB code for MCS, which may be found in the
function MedianCut.m under the IBL folder. The input for this function is
the HDR environment map using a latitude-longitude mapping stored in
img and the number of lights to be generated in nlights. The falloff
can be set off if the falloff in the environment map is pre-multiplied into
the input environment. This code initializes a set of global variables, and
the image is converted to luminance and stored in L. Other global variables are used to facilitate the computation. The function then calls the

MedianCutAux.m function, with the initial dividing axis along the longest
dimension. MedianCutAux.m may be found under the IBL/util folder and
represents the recursive part of the computation; see Listing 7.5. This
function computes the sum of luminance in the region and then identifies
the pivot point of where to split depending on the axis chosen.

```
function MedianCutAux(xMin, xMax, yMin, yMax, iter)
global L;
global imgWork;
global lights;

lx = xMax - xMin;
ly = yMax - yMin;

if((lx > 2) && (ly > 2) && (iter > 0))
    tot = sum(sum(L(yMin:yMax, xMin:xMax)));
    pivot = -1;

    if(lx > ly) %cut on the X-axis
        for i=xMin:(xMax - 1)
            c = sum(sum(L(yMin:yMax, xMin:i)));
            if(c >= (tot - c))
                pivot = i;
                break;
            end
        end

        if(pivot == -1)
            pivot = xMax-1;
        end

        MedianCutAux(xMin,    pivot, yMin, yMax, iter-1);
        MedianCutAux(pivot+1, xMax,  yMin, yMax, iter-1);
    else %cut on the Y-axis
        for i=yMin:(yMax - 1)
            c = sum(sum(L(yMin:i, xMin:xMax)));
            if(c >= (tot - c))
                pivot = i;
                break;
            end
        end

        if(pivot == -1)
            pivot = yMax-1;
        end

        MedianCutAux(xMin, xMax, yMin,    pivot, iter-1);
        MedianCutAux(xMin, xMax, pivot+1, yMax,  iter-1);
    end
else %create a light source
    lights = [lights, CreateLight(xMin, xMax, yMin, yMax, L,
        imgWork)];
```

(a)   (b)

Figure 7.7. MCS for IBL: (a) The input environment map. (b) A visualization of the cuts and samples for 32 samples.

```
end
end
```

Listing 7.5. MATLAB code: Recursive part of median cut for light source generation.

Finally, when the termination conditions are met, the light sources are generated based on the centroid of the computed regions using function CreateLight.m and stored into lights, assigning the average color of that region. The code for CreateLight.m is given in Listing 7.6 and may be found under the IBL/util folder.

```
function light = CreateLight(xMin, xMax, yMin, yMax, L, img)
tot  = (yMax - yMin + 1) * (xMax - xMin + 1);
tmpL = L(yMin:yMax, xMin:xMax);
totL = sum(tmpL(:));
if((tot > 0) && (totL > 0))
    %compute the light color value
    col = reshape(img(yMin:yMax, xMin:xMax,:), tot, 1, size(img
        , 3));
    value = sum(col,1);
    %compute the light position
    [r, c] = size(L);
    [X, Y] = meshgrid(xMin:xMax, yMin:yMax);
    x_light = sum(sum(tmpL .* X)) / (totL * c);
    y_light = sum(sum(tmpL .* Y)) / (totL * r);
    %adding the light source in the list
    light = struct('col', value, 'x', x_light, 'y', y_light);
else
    light = [];
end
end
```

Listing 7.6. MATLAB code: Generate light in the region for median cut algorithm.

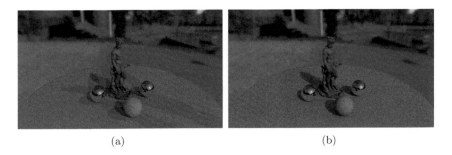

(a)   (b)

Figure 7.8. An example of evaluation of Equation (7.7) using MCS [108] with different N: (a) $N = 8$. Note that aliasing artifacts appear such as multiple shadows and incorrect shading. (b) $N = 256$; aliasing is alleviated.

After the generation of light sources, Equation (7.6) is evaluated as

$$L(\mathbf{x}, \boldsymbol{\omega}) = L_e + \sum_{i=1}^{N} C_i f_r(\boldsymbol{\omega}'_i, \boldsymbol{\omega})(\mathbf{n} \cdot \boldsymbol{\omega}'_i) V(\mathbf{x}, \boldsymbol{\omega}'_i), \qquad (7.7)$$

where N is the number of generated light sources, $-\boldsymbol{\omega}'_i$ is the direction of the generated light source, and C_i is the corresponding color. Figure 7.7 and Figure 7.8 demonstrate examples of the method.

Typically, these methods generate results with less noise, so they are ideal for animated scenes, where the geometry and camera can be dynamic but the environment map is a still image. However, rendered images can have aliasing artifacts if only a few light sources are generated; depending on the radiance distribution and the dynamic range of the environment map. Figure 7.8 shows an example of aliasing artifacts caused by the limited number of generated lights.

7.2.2 Monte-Carlo Integration and Importance Sampling

Another popular method for IBL is Monte-Carlo integration. This method uses random sampling for evaluating complex multidimensional integrals, as in the case of Equation (7.6). As an example, a one-dimensional function, $f(x)$, to be integrated over the domain $[a, b]$ is usually solved as

$$I_{ab} = \int_a^b f(x) dx = F(a) - F(b), \quad F'(x) = f(x).$$

However, it may not be possible to integrate $F(x)$ analytically as is the case for a normal distribution or if $f(x)$ is known only at a few points over the domain. In Monte-Carlo integration [312], integration is calculated by averaging the value of $f(x)$ in N points distributed over a domain, assuming Riemann integrals:

$$\hat{I}_{ab} = \frac{b-a}{N} \sum_{i=1}^{N} f(x_i), \qquad I_{ab} = \lim_{N \to +\infty} \frac{b-a}{N} \sum_{i=1}^{N} f(x_i), \qquad (7.8)$$

where $x_1, x_2, ..., x_N$ are random uniformly distributed points in $[a, b]$. To integrate a multidimensional function equidistant point grids are needed, and these can be very large (N^d). Here N is the number of points for a dimension and d is the number of dimensions of $f(x)$.

(a)   (b)

Figure 7.9. An example of Importance sampling Monte-Carlo integration for IBL using Pharr and Humphreys' method: (a) Importance sampling with 8 samples per pixel. (b) Importance sampling with 256 samples per pixel. (The three-dimensional model of *Nettuno* (Neptune) is courtesy of the Visual Computing Laboratory ISTI-CNR.)

The convergence in the Monte-Carlo integration (Equation (7.8)) is determined by variance, $\sigma \propto N^{-\frac{1}{2}}$, which means that N has to be squared to half the error. A technique that reduces variance is called *importance sampling*. Importance sampling solves the integral by taking points x_i that contribute more to the final result. This is achieved by using a probability density function $p(x)$ with a corresponding shape to $f(x)$:

$$\hat{I}_{ab} = \frac{1}{N} \sum_{i=1}^{N} \frac{f(x_i)}{p(x_i)}.$$

Note that the variance is still the same, but a good choice of $p(x)$ can make it arbitrarily low. The optimal case is when $p(x) = \frac{f(x)}{I_{ab}}$. To create samples, x_i, according to $p(x)$ the inversion method can be applied. This method calculates the cumulative distribution function $P(x)$ of $p(x)$; then samples, x_i, are generated as $x_i = P^{-1}(y_i)$ where $y_i \in [0, 1]$ is a uniformly distributed random number.

Importance sampling can be straightforwardly applied to the IBL problem, extending the problem to more than one dimension [310]. Good choices of $p(x)$ are the luminance of the environment map, $l(\omega')$, or the BRDF, $f_r(\omega, \omega')$, or a combination of both. An example of the evaluation of IBL using Monte-Carlo integration is shown in Figure 7.9. Monte-Carlo methods are unbiased; they converge to the real value of the integral, but they have the disadvantage of noise, which can be alleviated with importance sampling.

```
function [imgOut, samples] = ImportanceSampling(img, falloff,
    nSamples)
if(falloff)
    img = FallOffEnvMap(img);
end

%compute luminance channel
L = lum(img);
[r, c] = size(L);

%create 1D distributions for sampling
cDistr = [];
values = zeros(c, 1);
for i=1:c
    %1D Distribution
    tmpDistr = Create1DDistribution(L(:, i));
    cDistr = [cDistr, tmpDistr];
    values(i) = tmpDistr.maxCDF;
end
rDistr = Create1DDistribution(values);

samples = [];
imgOut = zeros(size(L));
pi22 = 2 * pi^2;
for i=1:nSamples
    %sample rDistr
    [x, pdf1] = Sampling1DDistribution(rDistr, rand());
    %sample cDistr
    [y, pdf2] = Sampling1DDistribution(cDistr(x), rand());
    %compute the direction
    angles = pi * [2 * x / c, y / r];
    vec = PolarVec3(angles(2), angles(1));
    pdf = (pdf1 * pdf2) / (pi22 * abs(sin(angles(1))));
    %create a sample
    sample = struct('dir', vec, 'x', x/c, 'y', y/r, 'col', img(
        y,x,:), 'pdf', pdf);
    samples = [samples, sample];
    imgOut(y, x) = imgOut(y, x) + 1;
```

```
end
end
```

Listing 7.7. MATLAB code: Importance sampling of the hemisphere using the Pharr and Humphreys' method.

Listing 7.7, which may be found in the `ImportanceSampling.m` function under the IBL folder, provides the MATLAB code for Pharr and Humphreys' importance sampling method [310] that uses the luminance values of the environment map for importance sampling. This method creates a cumulative distribution function (CDF) based on the luminance (computed in L) of each of the columns and a CDF based on each of these columns over the rows for the input environment map `img`. The code demonstrates the construction of the row and column CDFs stored in `rdistr` and `cdistr`, respectively. The generation of `nSamples` subsequently follows. For each sample, two random numbers are generated and used to obtain a column and row, effectively with a higher probability of sampling areas of high luminance. The code outputs both the samples and a map visualizing where the samples are placed in `imgOut`. It is important to note that within a typical rendering environment, such as Pharr and Humphreys' physically-based renderer [310], the creation of the CDFs is computed once before the rendering phase. In the rendering phase, a number of samples to the environment is generated whenever shading via the environment is required. The MATLAB code would only apply to one of these shading points. Figure 7.10 shows the results of running Pharr and Humphreys' importance sampling.

(a)   (b)

Figure 7.10. Pharr and Humphreys' importance sampling for IBL: (a) The input environment map. (b) A visualization of a chosen set of 256 samples; note that most of samples are correctly clustered near the sun.

The importance sampling method of Pharr and Humphreys, and other methods that exclusively sample the environment map, may not be ideal when computing illumination for specular surfaces, as the chosen samples are independent of the BRDF and the contribution in the chosen directions may not be ideal. Similarly, importance sampling of only the BRDF may result in significant contributions of the incident lighting being overlooked. Ideally, all terms of the rendering equation are considered. Multiple importance sampling [391] was the first technique to introduce sampling of more than a single term to computer graphics. It presented a generic method for importance sampling of multiple terms. Other methods have been focussed more specifically at importance sampling in the case of IBL. Burke et al. [80] presented bidirectional importance sampling (BIS). BIS used rejection sampling and sampling importance resampling (SIR) to obtain samples from the product of the environment map and the BRDF. Rejection sampling requires an unknown number of retries to generate the product samples. SIR does not require a number of retries so the number of samples generated can be bounded. SIR was concurrently presented by Talbot et al. [363] in the context of IBL to importance sample the product of the BRDF and the environment map, where an initial set of samples drawn from one of the distributions is subsequently weighted and resampled to account for the second term. If both the lighting and BRDF have high frequencies, SIR may not be ideal as it becomes difficult to obtain samples that are representative.

Wavelet importance sampling [96] also performed importance sampling of the BRDF and the luminance in the environment map by storing the lighting and BRDF as sparse Haar wavelets. This method uses precomputation for computing the wavelets and may require considerable memory for anything but low resolution lighting. This work was further extended [95] to remove such limitations by sampling the BRDF in real time and building a hierarchical representation of the BRDF. This allows the support of arbitrary BRDFs and complex materials such as procedural shaders. The product with lighting is computed by multiplying the lighting, represented as a mip-map with the BRDF hierarchy, enabling much higher resolution environment maps. Results showed how this method compared favorably with the previously discussed methods.

While these methods all account for two of the terms in the rendering equation, occlusion represented by V in Equation (7.6) is not taken into account. Clarberg and Akenine–Möller [94] used control variates to reduce the variance. This is done by approximating the occlusion using a visibility cache that provides a quick approximation of the lighting, which is in turn used to reduce variance.

The visibility problem is tackled by Bashford–Rogers et al. [56]. The sampling efficiency is improved by taking into account the view; a sampling distribution is automatically constructed in locations that are relevant to the camera position. To achieve this, the method performs two passes: a pre-pass, traces paths with the goal of creating a path map for direction selection onto the environment and a disk distribution for position selection.

(a)   (b)

Figure 7.11. An example of Bashford–Rogers et al.'s sampling method [56] in a challenging visibility situation: (a) A rendering split comparison: This method (up) converges faster (less noise) than classic methods (down) within the same computational time (these images are courtesy of Thomas Bashford–Rogers). (b) The input HDR environment map. (The image is courtesy of Jassim Happa.)

In the actual rendering pass, the position and direction for each ray are drawn by exploiting two distributions computed during the pre-pass; see Figure 7.11. The method is general and can be assimilated into already existing bidirectional integrators such as: bidirectional path-tracing [211], metropolis light transport [392], and progressive photon mapping [159]. A solution to the visibility problem is also provided by Bitterlie et al. [65], whereby rectangular portals are projected onto an axis aligned rectangle permitting more efficient sampling of visible regions.

7.2.3   PRT for Interactive IBL

A series of methods have been adopted to enable IBL to be used for interactive rendering. As mentioned earlier, environment maps have been used for rendering diffuse surfaces by filtering the environment maps. Ramamoorthi and Hanrahan [317] introduced a method to efficiently store an irradiance environment map representation by projecting it onto a basis function. At runtime the irradiance can be computed by evaluating this representation. Ramamoorthi and Hanrahan used spherical harmonic polynomials without needing to access a convolved environment map. This method did not take occlusion and interreflections into account but served to inspire a series of techniques that did. These were termed precomputed radiance transfer (PRT) [347] techniques. PRT, as the name implies, requires a precomputation stage that computes the lighting and the transfer components of rendering and then can compute the illumination in real time. These methods are suitable for interactive applications and have been adopted by the games industry, since they are very fast to compute.

This approach essentially requires the computation of inner products once the precomputation stage, which may be rather expensive, is finalized.

Assuming only diffuse surfaces entailing the BRDF, ρ is dependent only on $\mathbf{x}$, and the modified rendering equation used for IBL previously can be adjusted while ignoring L_e. Equation (7.6) then becomes:

$$L(\mathbf{x}, \boldsymbol{\omega}) = \frac{\rho}{\pi} \int_{\Omega} L(\boldsymbol{\omega}')V(\mathbf{x}, \boldsymbol{\omega}')(\mathbf{n} \cdot \boldsymbol{\omega}')d\boldsymbol{\omega}'.$$

PRT projects the lighting (L) and transfer functions (the rest of the integral) onto an orthonormal basis. $L(\boldsymbol{\omega}')$ can be approximated as:

$$L(\boldsymbol{\omega}') \approx \sum_k l_k y_k(\boldsymbol{\omega}'),$$

where y_k are the basis functions. In this case, spherical harmonics as used by the original PRT method are assumed, and l_k are the lighting coefficients computed as:

$$l_k = \int L(\boldsymbol{\omega}')y_k(\boldsymbol{\omega}')d\boldsymbol{\omega}'.$$

In this instance l_k is computed using Monte-Carlo integration, similar to the method described in the previous section. The transfer functions are similarly evaluated using Monte-Carlo integration based on:

$$t_k = \frac{\rho}{\pi} \int y_k(\boldsymbol{\omega}')V(\mathbf{x}, \omega')(\mathbf{n} \cdot \boldsymbol{\omega}')d\boldsymbol{\omega}', \tag{7.9}$$

where, in this case, the computation may be rather expensive as $V(\mathbf{x}, \omega')$ would need to be evaluated via ray casting.

The computation of the lighting coefficients, l_k, and transfer coefficients, t_k, represents the precomputation aspect of PRT. Once these are computed, the lighting can be evaluated as:

$$L(\mathbf{x}, \boldsymbol{\omega}) = \sum_k l_k t_k, \tag{7.10}$$

which is a straightforward dot product that can be computed, relatively efficiently, on current graphics hardware.

The example above would only compute direct lighting from the environment map, ignoring the light transport from secondary bounces required for global illumination. This can be added by considering the Neumann expansion:

$$L(\mathbf{x}, \boldsymbol{\omega}) = L_e(\mathbf{x}, \boldsymbol{\omega}) + L_1(\mathbf{x}, \boldsymbol{\omega}) + L_2(\mathbf{x}, \boldsymbol{\omega}) \dots,$$

where L_e is the light emitted and L_i is the lighting at $\mathbf{x}$ at the ith bounce. This leads to a global illumination version of Equation (7.10):

$$L(\mathbf{x}, \boldsymbol{\omega}) = \sum_k l_k(t_k^0 + t_k^1 + \ldots),$$

where t_k^0 is equivalent to t_k calculated in Equation (7.9), and t^i represents the transfer coefficients at the ith bounce; see Green [150] for further details.

The PRT described so far is limited to diffuse surfaces and low frequency lighting effects. Since the original publication, a number of researchers have sought to reduce the limitations of the original method. Ng et al. [287] used wavelets instead of spherical harmonics to include high frequency effects. Zhou et al. [436] compute dynamic shadows. Ng et al. [288] generalize the approach for all-frequency illumination for arbitrary BRDFs by decoupling visibility and materials, thus using three coefficients. Haar wavelets were used as basis functions. Further extensions have allowed light interreflections for dynamic scenes [185, 295]. Yue et al. [433] provided an extension focused on IBL for rendering interior scenes, where the key idea is to determine inside the model portals from which the light enters. This method provides high quality results, but these portals need to be manually placed by artists when modeling.

7.2.4   Rendering with More Dimensions

Adelson and Bergen [6] described the amount of radiance based on the seven-dimensional plenoptic function:

$$P(x, y, z, \theta, \phi, \lambda, t),$$

where (x, y, z) denotes the three-dimensional location at which the incident lighting is captured, (θ, ϕ) describe the direction, λ the wavelength of the light, and t the time.

The IBL demonstrated so far fixes (x, y, z) and t and would usually use three values for λ (red, green, and blue). Effectively, the previously described methods correspond to $P(\theta, \phi)$ for red, green, and blue. This entails that lighting is based on a single point, infinitely distant illumination, at one point in time, and it cannot capture lighting effects such as shadows, caustics, and shafts of light. Recently, research has begun to look into the IBL methods that take into account (x, y, z) and t.

Spatially varying IBL. Sato et al. [339] made use of two omnidirectional cameras to capture two environment maps corresponding to two spatial variations in the plenoptic function. They used stereo feature matching to construct a measured radiance distribution in the form of a triangular mesh, the vertices of which represent a light source. Similarly, Corsini et al. [98] proposed to capture two environment maps for each scene and to solve for spherical stereo [224]. In this case, the more traditional method of using two steel balls instead of omnidirectional cameras was used. When the geometry of the scene is extracted, omnidirectional light sources are generated for use in the three-dimensional scene. The omnidirectional light sources make this representation more amicable to modern many-light rendering methods, such as light cuts [398]. Figure 7.12 shows an example of Corsini et al.'s method.

Unger et al. [380] also calculated spatial variations in the plenoptic function. Their method, at the capture stage, densely generated a series of environment maps to create what they term an incident light field (ILF), after the light fields presented by Levoy and Hanrahan [223]. Unger et al. presented two capture methods. The first involved an array of mirror spheres and capturing the lighting incident on all these spheres. The second device consisted of a camera mounted onto a translational stage that would capture lighting at uniform positions along the stage. The captured ILF is then used for calculating the lighting inside a conventional ray tracing-based renderer. Whenever a ray hits the auxiliary geometry (typically a hemisphere) representing the location of the light field, the ray samples the ILF and bilinearly interpolates directionally and spatially between the corresponding captured environment maps. Unger et al. [377], subsequently extended this work, which took an infeasibly long time to capture the

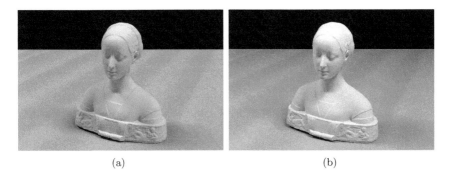

(a)   (b)

Figure 7.12. An example of stereo IBL by Corsini et al. [98] using the Laurana model: (a) The photograph of the original model. (b) The relit three-dimensional model of (a) using the stereo environment map technique. Note that local shadowing is preserved as in the photograph. (Images are courtesy of Massimiliano Corsini. The three-dimensional model of Laurana is courtesy of the Visual Computing Laboratory ISTI-CNR.)

Figure 7.13. An example of the dense sampling method by Unger et al. [377,380], the synthetic objects on the table are lit using around 50,000 HDR environment maps at Linköping Castle, Sweden. (Image is courtesy of Jonas Unger.)

lighting, by using the HDR video camera [376] described in Section 2.1.1. This method allowed the camera to roam freely, with the spatial location being maintained via motion tracking. The generated ILF consisted of a volume of thousands of light probes, and the authors presented methods for data reduction and editing. Monte-Carlo rendering techniques [310] were used for fast rendering of the ILF. Figure 7.13 shows an example when using this method.

A further improvement on ILF was introduced by Unger et al. [378,379]. They proposed a systems pipeline for capture, processing and rendering of so called *Virtual Photo Sets* (VPS). During capturing, HDR videos are used as input for image-based 3D reconstruction techniques; structure from motion (SfM) methods [361] with dense geometry estimation [361]. The obtained 3D geometry can be interactively adjusted with semi-automatic tools. Note that the output geometry can be improved by adding extra sensors such as laser scanners or depth cameras during the acquisition. At this point, lighting information from HDR videos is projected onto the 3D geometry and stored as 2D textures or 4D lightfields at the surface. From VPS, parameters of light sources, such as position, area, orientation, and color, can be estimated. Finally, the output VPS can be used for rendering photorealistic images. An example of this is shown in Figure 7.14.

Banterle et al. [42] introduced a system, *EnvyDepth*, that provides a stroke-based tool for estimating spatially varying illumination from a single HDR environment map. To capture local lighting effects, they proposed to decompose the incoming lighting in two components: a local component $L^l(y)$ where the light has a spatial component y, and a distant illumination $L^d(\omega)$ for far-away illumination; e.g., the sky. The resulting direct illumination computation can be written as:

$$L_o(x, n_x, \omega_o) = L_o^l(x, n_x, \omega_o) + L_o^d(x, n_x, \omega_o) = \quad (7.11)$$

$$\int_A L^l(y)V(x, \omega_{y,x})f_r(\omega_{y,x}, \omega_o)G(x, y)dA(y)+ \quad (7.12)$$

$$\int_\Omega L^d(\omega)V(x, \omega)f_r(\omega, \omega_o)(\omega \cdot n_x)d\omega \quad (7.13)$$

where L^l is the local component of real-world illumination, L^d the distant one, $G(x, y) = (\omega_{x,y} \cdot n_y)(\omega_{y,x} \cdot n_x)/\|x - y\|^2$ is the geometry term with $\omega_{x,y} = (x - y)/\|x - y\|^2$, and A is the surface of the scene. In this formulation, the contribution of L^d is computed as the classical distant lighting approximation, but only for the parts of the real-world scene that are, in fact, distant.

EnvyDepth is based on an iterative select-and-annotate workflow supported by a sketch-based interface. For each edit, the user selects regions of the environment map that should be reconstructed using a sketch-based metaphor [304]. Two types of strokes are placed directly on the environ-

(a)   (b)

Figure 7.14. An example of Virtual Photo Sets (VPS) rendering: (a) A virtual scene rendered using a traditional IBL using a single environment map. (b) The same scene rendered using the VPS method. The use of complex spatial and angular variations in the illumination enables an increase of realism level compared to standard IBL. (Images are courtesy of Jonas Unger.)

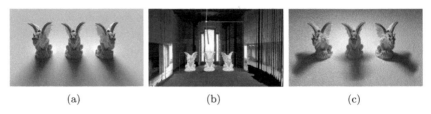

Figure 7.15. An example of *EnvyDepth*: (a) A rendering of three gargoyles using classic IBL and Paul Debevec's Ennis environment map. (b) A 3D reconstruction of the scene from the environment map used in (a). (c) A rendering of three gargoyles using the spatial lighting reconstruction in (b); note that shadows are spatially varying and are coherent with the nearby illumination from the window in (b). (The Ennis environment map is courtesy of Paul Debevec.)

ment map, to indicate regions that should be included or excluded from the selection. Then, edit propagation [19] is used to compute a selection from these imprecise strokes. At this point, the user assigns a geometric primitive for each selection such as: planes, which can be horizontal or vertical, extrusions from the profile of a plane, domes, and distant illumination. EnvyDepth outputs a depth map from which virtual point light sources are generated and used for rendering images in a straightforward way; see Figure 7.15.

Temporally varying IBL. As HDR video becomes more widespread, a number of methods that support lighting for IBL from dynamic environment maps, effectively corresponding to the change of t in the plenoptic function, have been developed. These methods take advantage of temporal coherence rather than recomputing the samples each frame, which may result in temporal noise.

Havran et al. [167] extended the static environment map importance sampling from their previous work [166] to be applicable in the temporal domain. Their method uses temporal filters to filter the power of the lights at each frame and the movement of the lights across frames. Wan et al. [399] introduced the spherical Q^2-tree. This is a hierarchical data structure that subdivides the environment map into equal quadrilaterals proportional to solid angles in the environment map. For static environment maps, the Q^2-tree creates a set of point lights based on the importance of the environment map in that area, similar to the light source generation methods presented in Section 7.2.1. When computing illumination due to a dynamic environment map, the given frame's Q^2-tree is constructed from that of the previous frame. The luminance of the current frame is inserted onto the Q^2-tree, which may result in inconsistencies since the Q^2-tree is based on the previous frame, so a number of merge and split operations update the

Q^2-tree until the process converges to that of a newly built Q^2-tree. However, to maintain coherence amongst frames and avoid temporal noise, the process can be terminated earlier based on a tolerance threshold.

Ghosh et al. [141] presented a method for sampling dynamically changing environment maps by extending the BIS method [80] into the temporal domain; see Section 7.2.2. This method supports product sampling of environment map luminance and BRDF over time. Sequential Monte-Carlo (SMC) was used for changing the weights of the samples of a distribution during consecutive frames. BIS was used for sampling in the initial frames. Resampling was used to reduce variance as the increase in number of frames could result in degeneration of the approximation. Furthermore, Metropolis-Hastings sampling was used for mutating the samples between frames to reduce variance. In the presented implementation, the samples were linked to a given pixel, and SMC was applied after each frame based on the previous pixel's samples. When the camera moved, the pixel samples were obtained by reprojecting the previous pixels' locations. Pixels without previous samples were computed using BIS.

Staton et al. [352] proposed to sample temporally varying environment maps by exploiting clustering. As a first step, their method generates importance samples for each frame as previously shown in Section 7.2.2. Then, these samples are clustered into representative light sources, which are shared across many frames of the video. K-means clustering is employed taking into account not only distance between clusters, but also luminance and time distance. Finally, a representative light for each cluster is created and it can be used for rendering.

(a)   (b)   (c)

Figure 7.16. An example of capturing a scene for differential rendering: (a) The photograph where to insert a synthetic object. (b) The photograph in (a) with markers for the camera calibration step. (c) An HDR environment map (tone mapped) encoding the lighting in (a).

7.2.5 Differential Rendering.

An important aspect when rendering synthetic objects with real-world lighting is to insert them seamlessly inside a photograph or video. In order to achieve this, Debevec [107] introduced a differential rendering methodology.

(a)   (b)

(c)   (d)

Figure 7.17. An example of inserting a synthetic object in a photograph (Figure 7.16): (a) The geometric corrected insertion of a synthetic object with local geometry. (b) The rendering of (a), I_{all}, using the HDR environment map, Figure 7.16(c). (c) The synthetic object mask, I_{mask}. (d) The rendering of the local geometry only., I_{local}.

The first step is to take a photograph of a scene, I, into which the synthetic object is to be inserted; see Figure 7.16(a). When taking this photograph, intrinsic and extrinsic camera values need to be calibrated. This can be achieved using markers [361]; see Figure 7.16(b). In order to maintain coherent lighting, an HDR environment map needs to be captured Figure 7.16(c). At this point, synthetic 3D models can be inserted with local geometry (i.e., a rough scene representation) for casting shadows; see Figure 7.17(a). However, rendering in a straightforward way and inserting the result in the photograph led to a seamy insertion because the local

(a)   (b)

Figure 7.18. An example of differential rendering: (a) The shadows to be transferred onto the photograph, I_{shadows}. (b) The final results; note that the insertion
is seamless.

geometry is still present. Therefore, there is the need to transfer shadows
onto the photograph and to remove the support geometry. This can be
achieved by the following steps:

- render the scene with the synthetic object with local geometry using the environment map as lighting, obtaining the image, I_{all}; see
  Figure 7.17(b);

- render a mask of only the synthetic object, I_{mask}, Figure 7.17(c);

- render the local geometry without the synthetic object, I_{local}, Figure 7.17(d).

Then, shadows to be transferred, see Figure 7.18(a) , can be computed
as:
$$I_{\text{shadows}} = \frac{I_{\text{all}} \times (1 - I_{\text{mask}})}{I_{\text{local}}} \tag{7.14}$$
and the final insertion is defined as:
$$I_{\text{out}} = I \times w(I_{\text{shadows}}) + I_{\text{all}} \times I_{\text{mask}} \tag{7.15}$$
where I_{out} is final result, and w is a function that changes black pixels
into white ones. Figure 7.18(b) shows the insertion of a synthetic object
in a seamless way. This methodology is very effective but it requires two
distinct renderings; I_{all} and I_{local}.

Virtual relighting. The plenoptic function considers only the capture of
fixed lighting at different positions, times, orientations, and wavelengths.
In order to have the ability of changing both the viewpoint and the lighting
a further number of factors, such as the location, orientation, wavelength,

Figure 7.19. An example of light stage: (a) A sample of four images from a database of captured light directions. (b) The relit scene captured in (a) using an environment map. (The Grace Cathedral environment map is courtesy of Paul Debevec.)

and timing of the light need to be considered. Debevec [109] considers this the reflectance field R, which is a 14-dimensional function accounting for both an incident ray of light L_i and the plenoptic function for reflected light L_r and given by:

$$R = R(L_i; L_r) = R(x_i, y_i, z_i, \theta_i, \phi_i, \lambda_i, t_i; x_r, y_r, z_r, \theta_r, \phi_r, \lambda_r, t_r),$$

where each term is equivalent to that in the plenoptic function for both the light and the view. When considering applications that make use of R, a number of approximations need to be taken into account. One popular application is the light stage and its various successors [110] (see Figure 7.19). These provide the ability of virtually relighting actors with arbitrary lighting after their performance is captured. The light stage captures the performance of an actor inside a rig surrounded by a multitude of lights.

The light rig lights the actor's face from each of the individual lights while the camera captures the actor's expression. The video capture accounts for θ_r and ϕ_r and the light rig for θ_i and ϕ_i. The view and light are considered static so (x_r, y_r, z_r) and (x_i, y_i, z_i) are constant. The time for the light taken to reach the actor's face is considered instantaneous, eliminating t_i. Similarly, the wavelength of light is not considered to be changing, removing λ_i, and fixing it to the three: red, green, and blue capture channels. The reflectance field can thus be approximated to $\tilde{R}(\theta_i, \phi_i, \theta_r, \phi_r, \lambda_r, t_r)$. The lights represent a basis function and which is subsequently used to relight the actor's face. The additive properties of light mean that subsequently the light values can be scaled by the contribution of the light representing that position in an environment map, giving the impression that the actor is lit from the environment map.

7.3 Summary

The widespread use of HDR has brought IBL to the forefront as one of its major applications. IBL has rapidly emerged as one of the most studied rendering methods and is now integrated in most rendering systems. The various techniques used at different ends of the computation spectrum have allowed simple methods such as environment mapping to be used extensively in games, while more advanced interactive methods such as PRT and its extensions begin to gain a strong following in such interactive environments. More advanced methods have been used and continue to be used in cinema and serious applications, including architecture and archaeology. As the potential of capturing more aspects of the plenoptic function (and indeed reflectance fields) increases, the ability to relight virtual scenes with real lighting will create many more possibilities and future applications.

HDR Images Compression

The extra information within an HDR image, compared to an LDR image, means that the resultant data files are large. Floating point representations, which were introduced in Chapter 2, can achieve a reduction down to 32/24 bpp (RGBE and LogLuv) from 96 bpp of an uncompressed HDR pixel. However, this memory reduction is not enough and not practical for easily distributing HDR content or storing large databases of images or video. For example, a minute of a high definition movie (1920×1080) at 24 fps encoded using the 24 bpp LogLuv format requires more than 8.3 GBytes of space, which is nearly double the space of a single layer DVD. Researchers have been working on more sophisticated compression schemes in the last few years to make storing of HDR content more practical. The main strategy has been to modify and/or adapt current compression standards and techniques such as JPEG, MPEG, and block truncation coding (BTC) to HDR content. This chapter presents a review of the state-of-the-art of these compression schemes for HDR images and textures, while the next chapter deals with HDR videos. The new JPEG supported standard for HDR is JPEG XT. This is expected to become a new standard for compressed HDR images and is also presented.

8.1 HDR Compression MATLAB Framework

The image compression framework consists of encoder and decoder functions used to compress HDR images. Once the HDR encoding function is completed, images are compressed using a standard LDR image encoder through the `imwrite.m` function of the Image Processing Toolbox of Mathworks [255]. When the compressed image needs to be decompressed, a decoder function decompresses it using the `imread.m` function of the Image Processing Toolbox and then it expands its range using the HDR decoding function and any meta-date required. In order to have a transparent framework, all operations are wrapped by two new IO functions: `hdrimread.m` for reading HDR images, and `hdrimwrite.m` for writing HDR images. Both functions can be found in the folder `IO`.

8.2   HDR Image Compression

This section introduces the main techniques for HDR image compression. Some of these concepts are used or extended to HDR texture and video compression. The overarching method for HDR compression is to reduce the dynamic range using tone mapping and to encode these images using standard encoding methods; see Figure 8.1. Subsequently, standard decoding and expansion operators are used for decoding. Additional information is stored to enable this subsequent expansion of the tone mapped images and to improve quality, including:

- Tone mapping parameters. These are the parameters of the range reduction function (which has an analytical inverse); they are needed to expand the signal back.

- Spatial inverse functions. These are the inverse tone mapping functions stored per pixel. These functions are obtained by dividing the HDR luminance channel by the tone mapped one. When they vary smoothly, depending on the TMO, they can be subsampled to increase efficiency.

- Residuals. These are usually the differences between the original HDR values and the reconstructed encoded values after quantization. The values significantly improve the quality of the final image because spatial quantization and bit reduction can introduce quanti-

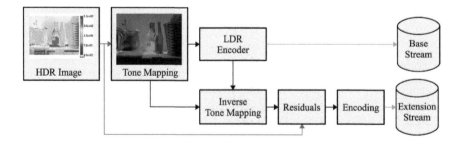

Figure 8.1. A scheme for generic HDR image compression.

zation errors. This can be noticed in the form of noise, enhancement of blocking and ringing, banding artifacts, etc.

The main differences between the various compression schemes are the choice of the LDR encoder, the way in which the range reduction function is calculated, and what data is stored to recover the full HDR image.

8.2.1 JPEG2000 and HDR

JPEG2000 is an efficient standard for encoding images based on the wavelet transform [134]. Although JPEG2000 Part 10 [136] introduced the ability to encode floating point data, not all implementations support it. Xu et al. [425] proposed a straightforward preprocessing technique that allows users to efficiently store HDR images using basic JPEG2000 implementations.

The encoding phase starts with the reduction of the dynamic range of an HDR image, I_w, by applying the natural logarithm to the RGB values:

$$I_{d,i}(\mathbf{x}) = \ln I_{w,i}(\mathbf{x}) \quad \forall i \in \{R, G, B\}.$$

Then, the floating-point values in the logarithmic domain are discretized to unsigned short integers ($n = 16$):

$$\overline{I}_{d,i}(\mathbf{x}) = (2^n - 1)\frac{I_{d,i}(\mathbf{x}) - I_{d,\,min,i}}{I_{d,\,max,i} - I_{d,\,min,i}} \quad \forall i \in \{R, G, B\}. \tag{8.1}$$

Finally, the image is compressed using a JPEG2000 encoder.

To decode, the image is first decompressed using a JPEG2000 decoder, then it is converted from integer into floating-point values by inverting Equation (8.1), which is subsequently exponentiated:

$$I_{w,i}(\mathbf{x}) = \exp\left(\frac{\overline{I}_{d,i}(\mathbf{x})}{2^n - 1}\left(I_{d,\,max,i} - I_{d,\,min,i}\right)\right) \quad \forall i \in \{R, G, B\}.$$

This method was compared in the JPEG2000 lossy mode against JPEG-HDR [410] and HDRV [245] and in JPEG2000 lossless mode against RGBE [405], LogLuv [214], and OpenEXR [180]. The metrics employed were RMSE in the logarithm domain, and Lubin's visual discrimination model [234]. The results of these comparisons showed that HDR-JPEG2000 in lossy mode is superior to JPEG-HDR and HDRV, especially at low bit rates when these methods have artifacts. Nevertheless, the method does not perform well when lossless JPEG2000 is used, because the file size is higher than the file size when using RGBE, LogLuv, and OpenEXR. Note that these methods are lossy in terms of the color representation because the color is quantized, but they do not apply spatial compression. Evaluation

showed that HDR-JPEG2000 can compress HDR images at 0.48-4.8 bpp
with

The HDR-JPEG2000 algorithm is a straightforward method for lossy
compression of HDR images at high quality without artifacts at low bit
rates. However, it inherits the same main issue of JPEG2000, i.e., it is not
a highly adopted standard.

The code for the encoder of HDR-JPEG2000 method is shown in Listing 8.1. The code of the encoder can be found in the file HDRJPEG2000

```
imgLog = log(img + 1e-6); %range reduction
col = size(img, 3);
xMin = zeros(col, 1);
xMax = zeros(col, 1);
metadata = [];
for i = 1:col
    xMin(i) = min(min(imgLog(:,:,i)));
    xMax(i) = max(max(imgLog(:,:,i)));
    delta = xMax(i) - xMin(i);
    imgLog(:,:,i) = (imgLog(:,:,i) - xMin(i)) / delta;
    metadata = [metadata, num2str(xMax(i)), '␣', num2str(xMin(i
        )), '␣'];
end
metadata = [metadata, num2str(nBit)];

imgLog = uint16(imgLog * (2^nBit - 1)); %quantization
imwrite(imgLog, name, 'CompressionRatio', compRatio, 'mode', '
    lossy', 'Comment', metadata);
```

Listing 8.1. MATLAB code: The HDR-JPEG2000 encoder implementing the
compression method by Xu et al. [425].

Enc.m under the folder Compression. The function takes as input the HDR
image to compress, img, the output name for the compressed image, name,
the compression ratio, compRatio, which has to be set greater than one,
and the quantization bits, nBit. As a first step, the function checks if inputs
are correct (missing from the listing). Subsequently, the image is stored
in the logarithmic domain, imgLog, and each color channel is separately
normalized in $[0, 1]$. Then, imgLog is stored into a JPEG2000 file using
the imwrite.m function (note that images can be saved in the JPEG2000
format only from MATLAB version 2010a) with metadata (xMin, xMax,
and nBit) as comments.

The code for decoding is shown in Listing 8.2. The code of the decoder
can be found in the file HDRJPEG2000Dec.m under the folder Compression.

```
info = imfinfo(name); %read metadata
```

```
decoded = sscanf(cell2mat(info.Comments), '%g', 7);
nBit = decoded(end);
imgRec = double(imread(name)) / (2^nBit - 1);
for i=1:size(imgRec, 3)
    xMax = decoded((i - 1) * 2 + 1);
    xMin = decoded((i - 1) * 2 + 2);
    %range expansion
    imgRec(:,:,i) = exp(imgRec(:,:,i) * (xMax - xMin) + xMin) -
        1e-6;
end
imgRec(imgRec < 0.0) = 0;
```

Listing 8.2. MATLAB code: The HDR-JPEG2000 decoder implementing the compression method by Xu et al. [425].

The function takes as input the name of the compressed image (without any file extension, i.e., similar input to the encoder). Note that the decoding process is quite straightforward and it just reverses the order of operations of the encoder.

8.2.2   JPEG XR

JPEG XR is a file format for compressing still images [135]. It is a standard (ISO/IEC 29199) and based on Microsoft's HD Photo. JPEG XR is a modern file format that supports images with an arbitrary number of color channels (monochrome, RGB, $CMYK$, and n channels images), different bit-depths (8, 16, and 32 bits per color channel), lossy and lossless compression (including RGBE for HDR images), different color representations, transparency (alpha channel), metadata, numbers representations (unsigned integer, fixed-point, and floating point), etc. Such capabilities mean that it can support HDR via the increased bit-depths using methods similar to those for JPEG2000, discussed above, and directly via half and full floating point support.

One of the most important differences between JPEG and JPEG XR is the core compression algorithm. JPEG employs the discrete cosine transform (DCT) on 8×8 image blocks to convert pixel values into the frequency domain, where they are discretized using quantization matrices. JPEG XR, on the other hand, uses a hierarchical two-stage lapped biorthogonal transform [370], which is invertible in integer arithmetic. Note that this transform can provide lossy and lossless compression in the same framework.

Although JPEG XR is very flexible and provides better quality and lower computational complexity than classic JPEG for encoding LDR images, the format adoption has not reached the popularity of JPEG because its innovative transform makes the standard not backward compatible.

8.2.3 JPEG XT

The rise in popularity of HDR has resulted in a large number of methods which are supported, for example in commercial cameras and camera phones. However, most of these are limited to vendor-lock formats, which makes it difficult to use these images widely. Furthermore, the results of these formats is an 8-bit LDR image and thus the original HDR content is lost.

Researchers have made a number of attempts to fit HDR images into the JPEG standard [51, 290, 409, 410]. However, these solutions have had limitations that make them less applicable as an industry standard; for example, Wide Color Gamut (WCG) compatibility, computational efficiency, etc.

In addition, industry is typically reluctant to change their technology pipeline in order to support a completely new standard. They prefer to adapt any new standard into the existing ecosystem keeping the necessary investment for the development of a new product relatively low.

To solve this issue, in 2012, the Joint Photographic Experts Group (JPEG), formally known as ISO/IEC JTC1/SC29/WG1, began the development of a new standard technology called JPEG XT (ISO/IEC 18477) [327]. The JPEG XT image coding system is organized into nine parts that define the baseline coding architecture (the legacy JPEG code stream 8-bit mode). This is an extensible file format specifying a common syntax for extending the legacy JPEG, and application of this syntax for coding integer or floating point samples within 8–16 bits precision [24]. This coding architecture is further refined to enable lossless and near-lossless coding, and it is complemented by an extension for representing alpha-channels. Thanks to its flexible layered structure, the JPEG XT capabilities can be extended into novel applications such as omni-directional photography, animated images, structural editing as well as privacy, and security. These are currently under examination and development [326].

In practice, JPEG XT can be seen as a superset of the 8-bit mode in which JPEG where existing JPEG technology is reused whenever possible. This allows users to decode an LDR image from a JPEG XT stream via legacy JPEG implementations [24]. The JPEG XT architecture has two layers. The first layer stores an LDR image as a JPEG image at 8-bits per sample in the ITU BT.601 RGB color space [381], i.e., the base layer b. The second layer, E, includes the additional information to reconstruct the HDR image I_w starting from b.

The standard specifies only the decoder level. A JPEG XT encoder receives as input a pair of images: an HDR and an LDR image. The LDR image is a tone mapped version of the HDR input image, which is stored with a legacy JPEG 8-bit codec. E can be computed in many ways.

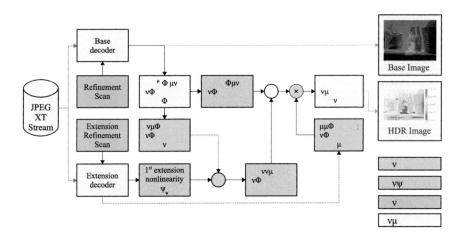

Figure 8.2. The decoding pipeline for the JPEG XT standard [24]. Note that dashed lines represent scalar values, and continue lines represent vector values.

Typically, this happens by either using a division/multiplication or subtraction operations between the LDR and the HDR image. Then, this may be followed by pre-processing steps to improve the encoding performances. Similar to JPEG, the pre-processed input is then decorrelated with a discrete cosine transform (DCT), quantized, and entropy coded [23]. A further mechanism, called refinement scan, was introduced to extend the coding precision in the DCT domain. This is closely related to the progressive coding mode of the legacy JPEG standard, and it extends the coding precision to 12-bits in the spatial domain.

Figure 8.2 shows the typical JPEG XT standard decoding workflow. Note that, not all the implementation of these components are necessary for properly reconstructing the HDR image from b and E layers. Profiles were introduced to identify a subset of these components to provide a suitable decoding procedure, and three main profiles were defined: A, B, and C. All profiles have in common the use of two 3×3 matrices operations meant for two different scopes. The first matrix, $\mathbf{C}$, implements a color transformation that takes into account the possibility to have an HDR input image expressed in a color space different from the one used to represent the LDR image, i.e., ITU BT.601 RGB. $\mathbf{C}$ converts from ITU-R BT.601 to the target colorspace in the extension layer. The second matrix, $\mathbf{R}$, implements the de-correlation transformation from $YCbCr$ to RGB in the extension layer to clearly separate the luminance component Y from the chromaticity ($CbCr$) at encoding level. Furthermore, a first base non-linearity transformation is common to all profiles. This is an inverse gamma function and applied to the base layer, b. However, profile C also implements an inverse global tone mapping function.

In Profile A, b is first inverse gamma corrected by applying the nonlinear function Φ, and then is multiplied by a luminance scale μ to reconstruct I_w as

$$I_w(\mathbf{x}) = \mu\big(E_l(\mathbf{x})\big)\Big[\mathbf{C} \cdot \Phi\big(b(\mathbf{x})\big) + \nu\Big(S\,\mathbf{C} \cdot \Phi\big(b(\mathbf{x})\big)\Big)\,\mathbf{R} \cdot E^{\perp}(\mathbf{x})\Big], \quad (8.2)$$

where μ is a scalar function of the luma component of the extension layer (E_l), and $E^{\perp}$ is the extension layer projected onto the chroma-subspace, i.e., E with its luma component set to zero. S is a row-vector transforming a color into a luminance value, and ν is a scalar function taking in input luminance values:

$$\nu(x) = x + \epsilon \qquad (8.3)$$

where ϵ is a *noise floor* that avoids instability in the encoder for very dark image regions.

The reconstruction of I_w, for Profile B, is based on the trick that a quotient can be expressed as sum in a logarithmic space:

$$I_w(\mathbf{x})_i = \sigma \exp\Big(\log\big(\big[\mathbf{C} \cdot \Phi\big(b(\mathbf{x})\big)\big]_i\big) - \log\big(\Psi\big(\big[\mathbf{R} \cdot E(\mathbf{x})\big]_i\big) + \epsilon\big)\Big),$$

where $i = \{R, G, B\}$ is the index of the RGB color channels. Here two inverse gamma functions Φ and Ψ are used. Φ linearizes the base layer, while Ψ is meant to better distribute values closer to zero in the extension layer at encoder level. σ is a scalar value used as an exposure parameter, which optimizes the split between base and extension layers.

In Profile C, I_w is reconstructed by adding the base and extension layers. To avoid additional transformation in decoding, the extension layer is encoded directly in logarithmic space:

$$I_w(\mathbf{x}) = \psi \exp\Big(\hat{\Phi}\big(\mathbf{C} \cdot b(\mathbf{x})\big) + \mathbf{R} \cdot E(\mathbf{x}) - 2^{15}(1, 1, 1)^{\top}\Big); \quad (8.4)$$

where $\hat{\Phi}(x) = \psi \log\big(\Phi(x)\big)$, Φ is the global inverse tone-mapping approximation stored as a lookup-table, and 2^{15} is an offset shift to make the extension image symmetric around zero [24].

Subjective and objective evaluations of this new standard, at the time of writing, were performed [23, 200, 249]. These showed that JPEG XT can encode HDR images at bit rates varying from 1.1 to 1.9 bpp for estimated mean opinion score (MOS) values above 4.5 out of 5, which is considered as fully transparent in many applications. Furthermore, JPEG XT can produce, on average, a bit stream that is 23 times smaller than the one created by lossless OpenEXR PIZ compression [23]. Recently, more extensive objective evaluations were performed [249] showing how different encoding tools can be combined to improve the overall image quality.

8.3   HDR Texture Compression

The focus of this section is on the compression of HDR textures, which are images used in computer graphics for increasing details of materials in a three-dimensional scene. The main difference between compressed textures and images is that the fetch of a pixel value has to happen in constant time allowing random access to the information. The drawback in having a random access is the limit in compression rates, around 8 bpp, because spatial redundancy is not fully exploited. Note that in the methods of the previous section, the full image needs to be decoded before providing access to the pixels, while with these texture methods this is not the case. Most of

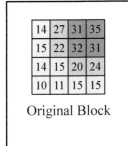
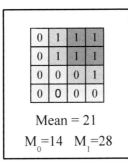
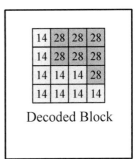

Figure 8.3. An example of BTC.

the texture methods are based on Block Truncation Coding (BTC) [170]. BTC is a compression scheme that divides the image in 4×4 pixels blocks. For each block the average value, m, is calculated and each pixel, x, is encoded as 0 if $x < m$, and as 1 if not. Then, the means of each group of pixels is calculated and stored. During the encoding, the mean of each group is assigned to their pixels; see Figure 8.3.

A typical BTC scheme for textures is S3TC by Iourcha et al. [181] and inspired by Knittel et al.'s work [197]; see Figure 8.4. This is called block compression (BC) in Microsoft Direct3D 11 and 12 (BC1, BC2, and BC3) and S3TC in OpenGL using the EXT_texture_compression_s3tc extension. This scheme projects pixel values of a 4×4 block into a line defined by two base colors, C_0 and C_1, which are typically computed by applying PCA to the block. The encoded values are: C_0 and C_1, which are quantized using 16 bits (5 bits for red channel, 6 bits for the green channel, and 5 bits for the blue channel), and each projected pixel value as a point (i.e., a color index), $p \in [0, 1]$, on the line C_0-C_1 using 2 bits. During the decoding, a pixel value is the result of linear interpolation between C_0 and C_1 at point p.

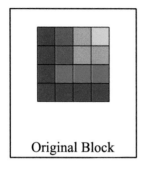 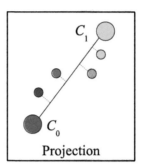 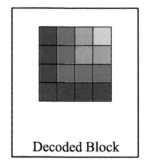

| Original Block | Projection | Decoded Block |

Figure 8.4. An example of S3TC.

The texture compression schemes in this section are a trade-off between compression rates, hardware support, speed of encoding, speed of decoding, and quality. The choice of the correct scheme depends on the constraints of the application.

8.3.1 Non-Backward Compatible Schemes

Researchers have proposed non-backward compatible schemes in order to achieve high quality texture compression at 8 bpp. Although, these schemes are not compatible with hardware implemented LDR schemes; e.g. S3TC, they share concepts common to texture compression such as 4 × 4 blocks compression, and random access.

HDR texture compression using geometry shapes. One of the first BTC schemes for HDR textures was proposed by Munkberg et al. [277]. This scheme compresses 48 bpp HDR textures into 8 bpp, leveraging on logarithmic encoding of luminance and geometry shape fitting for the chrominance channel.

The first step in the decoding scheme is to transform data into a color space where luminance and chrominance are separated to achieve a better compression. For this task, the authors defined a color space, $\overline{Y}\overline{u}\overline{v}$, as

$$\begin{bmatrix} \overline{Y}_{\mathrm{w}}(\mathbf{x}) \\ \overline{u}_{\mathrm{w}}(\mathbf{x}) \\ \overline{v}_{\mathrm{w}}(\mathbf{x}) \end{bmatrix} = \begin{bmatrix} \log_2 Y_{\mathrm{w}}(\mathbf{x}) \\ 0.114\frac{1}{Y_{\mathrm{w}}(\mathbf{x})}B_{\mathrm{w}}(\mathbf{x}) \\ 0.299\frac{1}{Y_{\mathrm{w}}(\mathbf{x})}R_{\mathrm{w}}(\mathbf{x}) \end{bmatrix}, \qquad Y_{\mathrm{w}}(\mathbf{x}) = \begin{bmatrix} 0.299 \\ 0.587 \\ 0.114 \end{bmatrix}^{\top} \cdot \begin{bmatrix} R_{\mathrm{w}}(\mathbf{x}) \\ G_{\mathrm{w}}(\mathbf{x}) \\ B_{\mathrm{w}}(\mathbf{x}) \end{bmatrix},$$

where $\overline{u}$ and $\overline{v}$ are in $[0,1]$, with $\overline{u} + \overline{v} \leq 1$. The image is then divided into 4 × 4 pixels blocks. For each block, the maximum, $\overline{Y}_{\max}$, and the minimum, $\overline{Y}_{\min}$, luminance values are calculated; see Table 8.1. These values are quantized at 8 bits and stored to be used as base luminance

byte	7	6	5	4	3	2	1	0
0	$\overline{Y}_{\text{max}}$							
1	$\overline{Y}_{\text{min}}$							
2	$\overline{Y}_0$			$\overline{Y}_1$			$\cdots$	
				$\cdots$				
10	$\text{type}_{\text{shape}}$				$\overline{u}_{\text{start}}$			
11	$\overline{u}_{\text{start}}$				$\overline{v}_{\text{start}}$			
12	$\overline{v}_{\text{start}}$			$\overline{u}_{\text{end}}$				
13	$\overline{u}_{\text{end}}$			$\overline{v}_{\text{end}}$				
14	ind_0		ind_1		ind_2		ind_3	
15	ind_4		ind_5		ind_6		ind_7	

Table 8.1. The table shows bit allocation for a 4×4 block in Munkberg et al.'s method [277].

values for the interpolation in a similar way to S3TC [181]. Moreover, the other luminance values are encoded with 2 bits, which minimize the value of the interpolation between $\overline{Y}_{\text{min}}$ and $\overline{Y}_{\text{max}}$.

At this point, chrominance values are compressed. The first step is to halve the resolution of the chrominance channel. For each block, a two-dimensional shape is chosen as the one that fits chrominance values in the $(\overline{u}, \overline{v})$ plane minimizing the error; see Figure 8.5. Finally, a 2-bit index is stored for each pixel that points to a sample along the fitted two-dimensional shape.

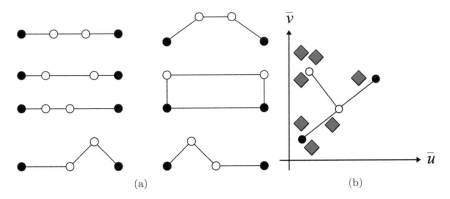

(a)   (b)

Figure 8.5. The encoding of chrominance in Munkberg et al. [277]: (a) Two-dimensional shapes used in the encoder. Black circles are for start and end, white circles are for interpolated values. (b) An example of a two-dimensional shape fitting for a chrominance block.

In the decoding scheme, luminance is firstly decompressed, interpolating values for each pixel in the block as

$$\overline{Y}_{\mathrm{w}}(\mathbf{y}) = \frac{1}{3}\left(\overline{Y}_{k(\mathbf{y})}\overline{Y}_{\min} + (1 - \overline{Y}_{k(\mathbf{y})})\overline{Y}_{\max}\right),$$

where $\overline{Y}_{k(\mathbf{y})}$ is the $\overline{Y}$ that corresponds to a pixel at location $\mathbf{y}$. The chrominance is then decoded as

$$\begin{bmatrix}\overline{u}_{\mathrm{w}}(\mathbf{y}) \\ \overline{v}_{\mathrm{w}}(\mathbf{y})\end{bmatrix} = \alpha(\mathrm{ind}_{k(\mathbf{y})})\begin{bmatrix}\overline{u}_{\mathrm{start}} - \overline{u}_{\mathrm{end}} \\ \overline{v}_{\mathrm{start}} - \overline{v}_{\mathrm{end}}\end{bmatrix} + \beta(\mathrm{ind}_{k(\mathbf{y})})\begin{bmatrix}\overline{v}_{\mathrm{end}} - \overline{v}_{\mathrm{start}} \\ \overline{u}_{\mathrm{start}} - \overline{u}_{\mathrm{end}}\end{bmatrix} + \begin{bmatrix}\overline{u}_{\mathrm{start}} \\ \overline{v}_{\mathrm{start}}\end{bmatrix},$$

where α and β are parameters specific for each two-dimensional shape. Subsequently, chrominance is up-sampled to the original size. Finally, the inverse $\overline{Y}\overline{u}\overline{v}$ color space transform is applied, obtaining the reconstructed pixel:

$$\begin{bmatrix}R_{\mathrm{w}}(\mathbf{x}) \\ G_{\mathrm{w}}(\mathbf{x}) \\ B_{\mathrm{w}}(\mathbf{x})\end{bmatrix} = 2^{\overline{Y}_{\mathrm{w}}(\mathbf{x})}\begin{bmatrix}0.229^{-1} \\ 0.587^{-1} \\ 0.114^{-1}\end{bmatrix}^{\top}\begin{bmatrix}\overline{v}_{\mathrm{w}}(\mathbf{x}) \\ 1 - \overline{v}_{\mathrm{w}}(\mathbf{x}) - \overline{u}_{\mathrm{w}}(\mathbf{x}) \\ \overline{u}_{\mathrm{w}}(\mathbf{x})\end{bmatrix}.$$

Munkberg et al. compared their scheme on a dataset of 16 HDR textures against two HDR S3TC variants using mPSNR [277], $\log_2[RGB]$ RMSE [425], and HDR-VDP [243,246]. The results showed that the method presents higher quality than S3TC variants, especially perceptually; achieving 8 bpp HDR texture compression. However, the decompression method needs special hardware, so it cannot be implemented in current graphics hardware. Moreover, the shape fitting can take up to an hour for a 1 Megapixel image, which limits the scheme to fixed content.

HDR texture compression using bit and integer operations. Concurrently with Munkberg et al. [277], Roimela et al. [334] presented an 8 bpp BTC scheme for compressing 48 bpp HDR textures, which was later improved in Roimela et al. [335].

The first step of the coding scheme is to convert data of the texture into a color space suitable for compression purposes. A computationally efficient color space was defined, which splits RGB colors into luminance and chromaticity. The luminance $I_{\mathrm{w}}(\mathbf{x})$ is defined as

$$I_{\mathrm{w}}(\mathbf{x}) = \frac{1}{4}R_{\mathrm{w}}(\mathbf{x}) + \frac{1}{2}G_{\mathrm{w}}(\mathbf{x}) + \frac{1}{4}B_{\mathrm{w}}(\mathbf{x})$$

and chromaticity $[r_Q, b_Q]^{\top}$ as

$$\begin{bmatrix}r_Q \\ b_Q\end{bmatrix} = \frac{1}{4I_{\mathrm{w}}(\mathbf{x})}\begin{bmatrix}R_{\mathrm{w}}(\mathbf{x}) \\ B_{\mathrm{w}}(\mathbf{x})\end{bmatrix}.$$

byte / bit	7	6	5	4	3	2	1	0
0	I_{bias}						n_{zero}	
1	n_{zero}		lum_0				lum_1	
2	lum_1			lum_2				
				...				
10			...			lum_{15}		
11	lum_{15}		r_{bias}					b_{bias}
12	b_{bias}			c_{zero}				
13	r_0			b_0			r_1	
14	r_1		b_1			r_2		b_2
15	b_2			r_3			b_3	

Table 8.2. The table shows bit allocation for a 4×4 block in Roimela et al.'s method [334].

Then, the image is divided into 4×4 pixels blocks. For each block, the luminance value with the smallest bit pattern is calculated, I_{min}, and its ten least significant bits are zeroed, giving I_{bias} (only 6 bits are stored). Subsequently, I_{bias} is subtracted bit by bit from all luminance values in the block:

$$bit(I'_w(\mathbf{y})) = bit(I_w(\mathbf{y})) - bit(I_{bias}),$$

where bit operator denotes the integer bit representation of a floating-point number and $\mathbf{y}$ is a pixel in the block being processed. Values $I'_w(\mathbf{y})$ share a number of leading zero bits that do not need to be stored. Therefore, they are counted in the largest $I'_w(\mathbf{y})$. The counter, n_{zero}, is clamped to seven and stored in 3 bits. At this point, the $n_{zero} + 1$ least important bits are removed from each $I'_w(\mathbf{y})$ in the block, obtaining $lum_w(\mathbf{y})$, which is rounded and stored as 5 bits. Chromaticity is now compressed. Firstly, the resolution of the chromaticity channels is halved. Second, the same compression scheme for luminance is applied to chromaticity, having two bias values at 6 bits, one for r_Q, $r_{Q,bias}$, and the other for b_Q, $b_{Q,bias}$. Furthermore, there is a common zero counter, c_{zero}, and final values are rounded in 4 bits. The number of bits for luminance and chromaticity channels are respectively 88 bits and 40 bits for a total of 128 bits or 8 bpp. Table 8.2 shows the complete allocation of bits.

The first decoding step is to reconstruct luminance by bit shifting to the left each $lum_y(\mathbf{w})$ value $n_{zero} + 1$ times, and adding I_{bias}. Secondly, this operation is repeated for the chromaticity channel, which is subsequently up-sampled to the original size. Finally, the image is converted from Ir_Qb_Q

color space to RGB applying the inverse transform:

$$\begin{bmatrix} R_w(\mathbf{x}) \\ G_w(\mathbf{x}) \\ B_w(\mathbf{x}) \end{bmatrix} = I_w(\mathbf{x}) \begin{bmatrix} 4 & 0 & 0 \\ 0 & 2 & 0 \\ 0 & 0 & 4 \end{bmatrix} \begin{bmatrix} r_{Q,w}(\mathbf{x}) \\ 1 - r_{Q,w}(\mathbf{x}) - b_{Q,w}(\mathbf{x}) \\ b_{Q,w}(\mathbf{x}) \end{bmatrix}.$$

This scheme was compared against Munkberg et al.'s scheme [277], HDR-JPEG 2000 [425], and a HDR S3TC variant using different metrics: PSNR, mPSNR [277], HDR-VDP [243, 246], and RMSE. A data set of 18 HDR textures was tested. The results showed that the encoding method has quality similar to RGBE. Moreover, it is similar to Munkberg et al.'s scheme [277], but the chromaticity quality is lower.

This compression method presents a computationally efficient encoding/decoding scheme for 48 bpp HDR textures. Only integer and bit operations are needed. Furthermore, it achieves high quality images at only 8 bpp. However, the main drawback of the scheme is that it cannot be implemented on current graphics hardware.

DHTC: an effective DXTC-based HDR texture compression scheme. Sun et al. [359] presented a S3TC-based HDR texture compression (DHTC), which separately compresses luminance and color values. The scheme works on 4×4 pixel blocks, the same as previous block truncation compression methods.

The encoding starts with the separation of luminance and color information using a classic YUV color space which is inspired by Munkberg et al. [277], and defined as

$$Y = \sum_i w_i C_i \qquad S_i = \frac{C_i w_i}{Y}, \tag{8.5}$$

where Y values are clamped in $[2^{-15}, 2^{16}]$ and S_i in $[2^{-11}, 1]$. This color transformation allows only three parameters to be saved, Y, $U = S_r$, and $V = S_g$, because S_b can be reconstructed from previous ones. However, the blue channel can suffer from quantization error if it contains a small value. This problem is solved by encoding the smallest values leaving the largest for reconstruction. To save memory, the largest channel, Ch, is calculated per block:

$$\text{Ch} = \text{argmax}_{j \in (r,g,b)} \sum_{i \in \text{block}} S_i^j, \tag{8.6}$$

and it is stored using 4 bits in the block; see Table 8.3.

After color transformation, Y values are quantized into integer values in

byte / **bit**	7	6	5	4	3	2	1	0
0	L_0						$L1$	
1	L_1		T_{indx}				Ch	
2	M_0		M_1				M_2	
				...				
7	M_{13}		M_{14}				M_{15}	

Table 8.3. The table shows bits allocation for luminance values L_0 and L_1, and the modifier table M_0-M_{15} in a 4×4 block in Sun et al.'s method [359].

the logarithmic space using maximum and minimum values of each block:

$$Y_{int} = \left\lfloor \frac{\log_2 Y - L_0}{L_1 - L_0} \right\rfloor \tag{8.7}$$

$$L_0 = \log_2 \max(Y_i \in \text{block}) \qquad L_1 = \log_2 \min(Y_i \in \text{block})$$

where L_0 and L_1 are discretized at 5-bit and stored as block information; see Table 8.3. Finally, Y_{int} and UV are quantized at 8-bit. Note that UV can be adaptively stored in the linear or logarithmic domain per block to improve efficiency.

A further transformation is applied to improve efficiency during DXTC compression, the point translation transformation (PTT). This is obtained by forcing color values of a block to be in a line segment between two colors (U_0, Y_0, V_0) and (U_1, Y_1, V_1). The PTT was designed off-line as

$$m(M_{indx}, T_{indx}) = (-1)^{M_{indx}\&1} 2^{T_{indx}>>2}(1 + (T_{indx}\&3 + 1)(M_{indx} >> 1)) \tag{8.8}$$

$$Y_t = \left[Y_{int} + m(M_{indx}, T_{indx})\right]_0^{255}$$

where $T_{indx} \in [0, 15]$ is a parameter per block, and $M_{indx} \in [0, 15]$ are parameters per pixel; see Table 8.3. These values are calculated during compression using a minimization process. Finally, values are ready to be compressed using the standard DXT1 compression scheme; see Table 8.4.

The decoding stage is divided into a few steps. When a pixel of a block needs to be decoded, firstly the YUV color is reconstructed using a DXT1 decoder, then $m(M_{indx}, T_{indx})$, in Equation (8.8), is removed from the Y coordinate, which is converted from local luminance to global luminance inverting Equation (8.3.1). Finally, Y values are exponentiated (including UV if they were encoded in the logarithm space), and the RGB color values are obtained by inverting Equation (8.5).

The method was compared with Munkberg et al. [277], Roimela et al. [334], and Wang et al. [401] using mPSNR [277], $\log_2[RGB]$ RMSE [425],

byte / **bit**	7	6	5	4	3	2	1	0
0	U_0				Y_0			
1	Y_0			V_0				
2	U_1				Y_1			
3	Y_1			V_1				
4	C_0		C_1		C_2		C_3	
					...			
7	C_{12}		C_{13}		C_{14}		C_{15}	

Table 8.4. The table shows local color bits allocation for a 4×4 block in Sun et al.'s method [359].

and HDR-VDP ($P(X) = 0.75$) [243, 246] as quality metrics. On a dataset of 6 images, the method was shown to be slightly better than Munkberg et al. [277], which was the best performer of the compared methods.

This method can be used for encoding LDR textures with an alpha channel. The comparison of this using BC3/DXT5 [268] showed a slightly better quality for the PSNR metric on a dataset of 12 images.

Sun et al. [360] improved DHTC quality (from 45.2 mPSNR dB to 45.3 mPSNR dB) using relative values stored in m instead of absolute ones (Equation (8.8)). They also added a 4 bpp texture format with a reasonable quality (38 mPSNR dB) for low memory budgets and created a unified layered framework. This format achieves 4 bpp by merging four neighboring 4×4 blocks into an 8×8 block and by subsampling the chrominance channels.

8.3.2   Backward Compatible Schemes

Methods, such as the one of Munkberg et al. [277] and Roimela et al. [334, 335], require hardware modifications for fetching luminance and chroma efficiently, such as a decompressor unit. This section describes a few methods that can be implemented on current and old graphics cards. Although these methods may not optimal for current desktop applications, they may still be useful for mobile world, where current APIs still do not fully support HDR textures.

HDR texture compression encoding LDR and HDR parts. Wang et al. [401] proposed a compression method based on S3TC [181]. The main concept of this method is to split the HDR image into two parts: one part consisting of LDR values and the second part of HDR values; see Figure 8.6. The two parts are stored on two S3TC textures for a total of 16 bpp.

The encoding starts by splitting luminance and chrominance using the

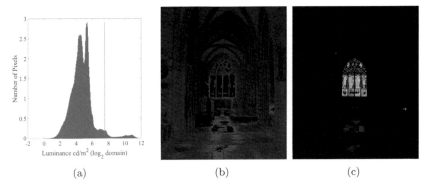

Figure 8.6. An example of the separation process of the LDR and HDR part in Wang et al. [401]: (a) The histogram of the image; the axis that divides the image in LDR and HDR is shown in red. (b) The LDR part of the image, uniformly quantized. (c) The HDR part of the image, uniformly quantized.

$LUVW$ color space, which is defined as

$$L_{\mathrm{w}}(\mathbf{x}) = \sqrt{R_{\mathrm{w}}(\mathbf{x})^2 + G_{\mathrm{w}}(\mathbf{x})^2 + B_{\mathrm{w}}(\mathbf{x})^2}, \quad \begin{bmatrix} U_{\mathrm{w}}(\mathbf{x}) \\ V_{\mathrm{w}}(\mathbf{x}) \\ W_{\mathrm{w}}(\mathbf{x}) \end{bmatrix} = \frac{1}{L_{\mathrm{w}}(\mathbf{x})} \begin{bmatrix} R_{\mathrm{w}}(\mathbf{x}) \\ G_{\mathrm{w}}(\mathbf{x}) \\ B_{\mathrm{w}}(\mathbf{x}) \end{bmatrix}.$$

After the color conversion, the luminance channel is split into an HDR and an LDR part. This is achieved by finding the value, $L_{\mathrm{w,\,s}}$, that minimizes the quantization error of encoding uniformly within the LDR and HDR parts separately. This is defined as

$$E(L_{\mathrm{w,\,s}}) = n_{\mathrm{LDR}} \frac{(L_{\mathrm{w,\,s}} - L_{\mathrm{w,\,min}})}{2^{b_{\mathrm{LDR}}}} + n_{\mathrm{HDR}} \frac{(L_{\mathrm{w,\,max}} - L_{\mathrm{w,\,s}})}{2^{b_{\mathrm{HDR}}}},$$

where n_{LDR} and n_{HDR} are respectively the number of pixels in the LDR and HDR part (both depend on $L_{\mathrm{w,\,s}}$). b_{LDR} and b_{HDR} are, respectively, the number of bits for quantizing the HDR and LDR parts. The HDR texture is stored in two S3TC textures, Tex_0 and Tex_1, as

$$\mathrm{Tex0}_{\mathrm{R}}(\mathbf{x}) = U_{\mathrm{w}}(\mathbf{x}), \qquad \mathrm{Tex0}_{\mathrm{G}}(\mathbf{x}) = V_{\mathrm{w}}(\mathbf{x}), \qquad \mathrm{Tex0}_{\mathrm{B}}(\mathbf{x}) = W_{\mathrm{w}}(\mathbf{x}),$$

$$\mathrm{Tex0}_{\mathrm{A}}(\mathbf{x}) = \begin{cases} \frac{1}{L_{\mathrm{w,\,s}} - L_{\mathrm{w,\,min}}} L_{\mathrm{w}}(\mathbf{x}) & \text{if } L_{\mathrm{w,\,min}} \le L_{\mathrm{w}}(\mathbf{x}) \le L_{\mathrm{w,\,s}}, \\ 0 & \text{otherwise}; \end{cases}$$

$$\mathrm{Tex1}_{\mathrm{A}}(\mathbf{x}) = \begin{cases} \frac{1}{L_{\mathrm{w,\,max}} - L_{\mathrm{w,\,s}}} L_{\mathrm{w}}(\mathbf{x}) & \text{if } L_{\mathrm{w,\,s}} < L_{\mathrm{w}}(\mathbf{x}) \le L_{\mathrm{w,\,max}}, \\ 1 & \text{otherwise}; \end{cases}$$

where the subscripts R, G, B, and A respectively indicate the red, green, blue, and alpha channels of a texture. Furthermore, additive residuals are included to improve quality. These are simply calculated as

$$\text{res}(\mathbf{x}) = L_{\text{w}}(\mathbf{x}) - L_{\text{w}}(\mathbf{x})',$$

where $L_{\text{w}}(\mathbf{x})'$ is the reconstructed luminance, which is defined as

$$L_{\text{w}}(\mathbf{x})' = \text{Tex0}_A(\mathbf{x})(L_{\text{w, max}} - L_{\text{w, s}}(\mathbf{x}))+$$
$$\text{Tex1}_A(\mathbf{x})(L_{\text{w, s}}(\mathbf{x}) - L_{\text{w, min}}) + L_{\text{w, min}}.$$

Then, res is split into three parts in a similar way to the luminance and stored in the red, green, and blue channels of Tex_1.

First, S3TC textures are decoded, then the luminance channel is reconstructed as

$$L_{\text{w}} = L_{\text{w}}(\mathbf{x})' + \text{Tex1}_R(\mathbf{x})(\text{res}_{s1} - \text{res}_{\text{min}}) + \text{Tex1}_G(\mathbf{x})(\text{res}_{s2} - \text{res}_{s1})+$$
$$\text{Tex1}_B(\mathbf{x})(\text{res}_{\text{max}} - \text{res}_{s2}) + \text{res}_{\text{min}}. \tag{8.9}$$

Finally, the image is converted from $LUVW$ color space to RGB color space:

$$\begin{bmatrix} R_{\text{w}}(\mathbf{x}) \\ G_{\text{w}}(\mathbf{x}) \\ B_{\text{w}}(\mathbf{x}) \end{bmatrix} = L_{\text{w}}(\mathbf{x}) \begin{bmatrix} U_{\text{w}}(\mathbf{x}) \\ V_{\text{w}}(\mathbf{x}) \\ W_{\text{w}}(\mathbf{x}) \end{bmatrix}.$$

The compression scheme was compared against RGBE and OpenEXR formats using classic PSNR, and the quality was worse than RGBE (5–7 dB less) and OpenEXR (40–60 dB less). Nevertheless, the method needs only 16 bpp compared with the 32 bpp used by RGBE and 48 used by OpenEXR. This compression scheme presents an acceptable quality and, since only simple operations are used in Equation (8.9), it can be mapped on current graphics hardware at high frame rates (e.g., 470–480 fps with a 512^2 viewport and texturing for each pixel). The main drawback of the method is that an optimal $L_{\text{w, s}}$ for an image can generate quantization artifacts that are noticeable.

HDR texture compression with tone mapping and its analytic inverse. Banterle et al. [45] presented a compression scheme for textures, which was designed to directly take advantage of graphics hardware. The generalized framework presented use a minimization process that takes into account the compression scheme for tone mapped images and residuals. Furthermore, the authors showed that up-sampling of tone mapped values before expansion does not introduce visible errors.

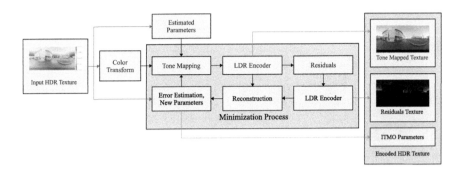

Figure 8.7. The encoding pipeline for Banterle et al.'s method [45].

The global Reinhard et al. operator [323] and its inverse [48] were employed in the implementation of the scheme. The forward operator is defined as

$$\begin{cases} f(L_{\mathrm{w}}(\mathbf{x})) = L_{\mathrm{d}}(\mathbf{x}) = \dfrac{\alpha L_{\mathrm{w}}(\mathbf{x})(\alpha L_{\mathrm{w}}(\mathbf{x}) + L_{\mathrm{w,\,H}} L_{\mathrm{white}}^2)}{L_{\mathrm{w,\,H}} L_{\mathrm{white}}^2 (\alpha L_{\mathrm{w}}(\mathbf{x}) + L_{\mathrm{w,\,H}})}, \\[2ex] \left[R_{\mathrm{d}}(\mathbf{x}), G_{\mathrm{d}}(\mathbf{x}), B_{\mathrm{d}}(\mathbf{x}) \right]^{\top} = \dfrac{L_{\mathrm{d}}(\mathbf{x})}{L_{\mathrm{w}}(\mathbf{x})} \left[R_{\mathrm{w}}(\mathbf{x}), G_{\mathrm{w}}(\mathbf{x}), B_{\mathrm{w}}(\mathbf{x}) \right]^{\top}, \end{cases}$$

where L_{white} is the luminance white point, $L_{\mathrm{w,\,H}}$ is the logarithmic average, and α is the scale factor. While the inverse is defined by

$$\begin{cases} g(L_{\mathrm{d}}(\mathbf{x})) = f^{-1}(L_{\mathrm{d}}(\mathbf{x})) = L_{\mathrm{w}}(\mathbf{x}), \\[1ex] \qquad = \dfrac{L_{\mathrm{white}}^2 L_{\mathrm{w,\,H}}}{2\alpha} \left(L_{\mathrm{d}}(\mathbf{x}) - 1 + \sqrt{(1 - L_{\mathrm{d}}(\mathbf{x}))^2 + \dfrac{4 L_{\mathrm{d}}(\mathbf{x})}{L_{\mathrm{white}}^2}} \right), \\[3ex] \left[R_{\mathrm{w}}(\mathbf{x}), G_{\mathrm{w}}(\mathbf{x}), B_{\mathrm{w}}(\mathbf{x}) \right]^{\top} = \dfrac{L_{\mathrm{w}}(\mathbf{x})}{L_{\mathrm{d}}(\mathbf{x})} \left[R_{\mathrm{d}}(\mathbf{x}), G_{\mathrm{d}}(\mathbf{x}), B_{\mathrm{d}}(\mathbf{x}) \right]^{\top}. \end{cases}$$

The first stage of encoding is to estimate parameters of the TMO, similarly to Reinhard's method [321], and to apply a color transformation to separate luminance and chrominance. Figure 8.7 shows the encoding pipeline. However, the color transformation can be skipped because S3TC does not support color spaces with separated luminance and chromaticity. Subsequently, the HDR texture and estimated tone mapping parameters are used as input in a Levenberg-Marquadt minimization loop, which ends when a local optimum for TMO parameters is reached. In the loop, the HDR texture is firstly tone mapped and encoded with S3TC. Then, residuals are calculated as differences and encoded using S3TC. To compute the error of the loop, the HDR texture is reconstructed. When a local optimum is reached, the HDR texture is tone mapped with the parameters and encoded using S3TC with residuals in the alpha channel.

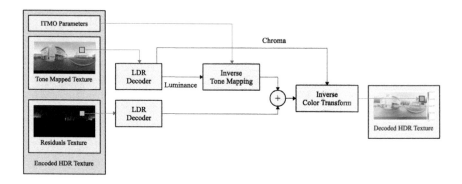

Figure 8.8. The decoding pipeline for Banterle et al.'s method [45].

The decoding stage can be implemented in a simple shader on a GPU; Figure 8.8 shows the decoding pipeline. When a texel is needed in a shader, the tone mapped texture is fetched and its luminance is calculated. The inverse tone mapping uses these luminance values, combined with the TMO parameters, to obtain the expanded values that are then added to the residuals. Finally, luminance and colors are recombined. Note that the inverse operator can be precomputed into a one-dimensional texture to speed up the decoding. Moreover, computations can be sped up by applying filtering during the fetch of the tone mapped texture. This is because the filtering is applied to coefficients of a polynomial function.

This method was compared to RGBE [405], Munkberg et al.'s method [277], Roimela et al.'s scheme [334], and Wang et al.'s scheme [402] using HDR-VDP [243], mPSNR [277], and RMSE in the logarithm domain [425]. The results showed that the approach presents a good trade-off between quality and compression, as well as the ability to decode textures in real time. Furthermore, it has a higher quality on average than Wang et al.'s method and it does not present contouring artifacts.

8.3.3   Hardware GPU Texture Compression Schemes

This section describes HDR texture compression methods that are currently implemented in modern GPUs and available in graphics' APIs such as OpenGL and Direct3D.

BC6H: This format, proposed by Microsoft [266], provides a high quality hardware compression scheme for HDR textures that are stored in half-precision (see Section 2.2) for both Direct3D (since version 11) and OpenGL (as ARB_texture_compression_bptc extension). BC6H shares similarity with S3TC: it works on 4×4 blocks with a budget of 128 bits for encoding a block, i.e., 8 bpp, and it has the base colors interpolation method. However, BC6H introduces three main novelties for handling HDR pixels. The first novelty is that the base colors, C_0 and C_1, can be stored as a color for C_0 (at high precision) and an offset d (at low precision) such that $C_1 = C_0 + d$, i.e., delta coding. This allows the encoder to improve quality especially when pixel values in a block are similar.

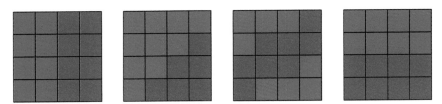

Figure 8.9. A sample (the other cases are similar) of the 32 two-regions separation's technique for BC6H (region 1 is red, and region 2 is blue).

The second novelty is the possibility of dividing the block into two regions (with 32 prefixed partitions), each of which has two base colors; see Figure 8.9. This feature enables the scheme to represent very difficult scenarios where there are more than two dominant colors, which are common when dealing with HDR content. Finally, the third and last novelty is the presence of 14 modes for encoding a block where each mode determines: different bits' allocations (base colors and pixel interpolation indices), the possibility to have delta coding, and the two-regions partition.

Although the method provides high quality compression and fast decoding on the GPU, the encoder is computationally very slow. Hence, a single block needs to be compressed with all possible modes and partitions to choose the best one. Given that there are 10 modes that allow the two-regions partition, 32 in total, a block needs to be compressed at least 324 times. Moreover, this scheme does not allow encoding of the alpha channel.

ASTC: This format was proposed by Nystad et al. [289] and provides compression in hardware for both LDR, 3D, and HDR textures. It is also available in both Direct3D (since version 12) and OpenGL (as KHR_texture _compression_astc_hdr extension). The novelty of this format is the ability to encode blocks of different size from 4×4 to 12×12 that can be rectangular with non power of two sizes. This allows the format to achieve different bit rates from 8 bpp (for 4×4 blocks) down to 0.89 bpp (for 12×12 blocks).

As with BC6H, ASTC has partitions (up to 4) that are procedurally computed and, as with S3TC, encodes values such that they are belonging to a line defined by two base colors, C_0 and C_1. For handling HDR values, the interpolation is a piecewise linear approximation to logarithmic interpolation. Furthermore, base colors are encoded as 12-bit unsigned floating point values (5 bits for the exponent and 7 bits for the mantissa). The final encoding can either use the delta coding (as in BC6H) or the scale coding where the second base color is encoded as $C_1 = C_0 \cdot s$, which can increase efficiency when C_1 is along the same luminance axis of C_0. Finally, both base colors and color interpolation indices are represented as *bounded integer sequence encoding* that allows the encoder to use a fractional number of bits increasing its flexibility.

The authors tested ASTC against BC6H for HDR textures compression showing that they have similar performances; BC6H has a slightly better quality [289]. However, the strength of this scheme lies in its ability to target different memory budgets, from 8 bpp down to 0.89 bpp, that makes it very flexible.

Method	bpp	St	BC
IMAGE COMPRESSION			
HDR-JPEG2000 [425]	1.6-3.2[1]	1	no
JPEG XR [135]	0.5-2[2]	1	no
JPEG XT [24]	1.1-1.9[3]	1[+]	yes
TEXTURE COMPRESSION			
Munkberg et al. [277]	8	1	No
Roimela et al. [334]	8	1	No
Wang et al. [401][H]	16	2	yes
Banterle et al. [45][H]	4-8	$1 + \alpha$	yes
Sun et al. [360]	4-8	1	no
BC6H [266][H]	8	1	no
Nystad et al. [289][H]	0.89-8	1	no

Table 8.5. Summary of the various image and texture compression methods. **bpp** refers to average bit per pixel for compressing an image/texture. **St** refers to single or dual images but does not include any metadata. **BC** refers to backwards compatibility with legacy LDR encoders in which one image/texture will output a viewable LDR image/texture. [H] means hardware support in the case of textures. [+] means that, depending on the profile, an extra layer may be required. [1] means that bpp range is achieved evaluating the scheme using the log $RMSE$ metric [51,425]. [2] means that bpp range is achieved evaluating the scheme using mean relative squared error (MRSE) [325]. [3] means that bpp range is achieved evaluating the scheme using HDR-VDP 2 (see Section 10.1.2) and subjective studies.

8.4 Summary

In the worst case, HDR still images for imaging and computer graphics applications requires 12 bytes per pixel for RGB images with floating point encoding. Recently, researchers and the industry have managed to provide high quality and efficient standards for every day needs. For example, JPEG XT represents a significant achievement for a wide range of applications and especially photography ensuring backward compatibility with the classic JPEG format. Table 8.5 summarizes what has been achieved to-date for compressing HDR images and textures. Note that texture compression methods usually have a higher bpp than image compression methods because textures have the main requirement to have a fast random access, which sacrifices the bpp. Furthermore, hardware schemes for computer graphics applications (e.g., virtual reality) have started to appear providing efficient texture compression formats.

HDR Video Compression

HDR video compression presents several similarities with HDR image compression. Many methods use range compression via tone mapping while maintaining residuals. All methods use LDR standards for encoding, although via the use of higher bit-depths and/or multiple streams. Some of the methods also employ coherence in between frames to avoid temporal artifacts. This chapter presents the methods which were originally designed with HDR video compression in mind.

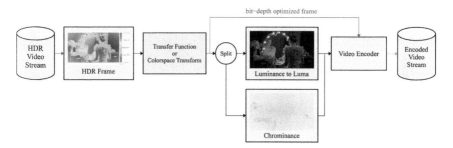

Figure 9.1. The encoding pipeline for single stream methods. (The original HDR video is courtesy of Jan Fröhlich.)

The HDR video compression methods can be broadly divided into two categories: single stream methods and dual stream methods. Single stream methods use higher bit-depths ($\geq$ 10 bits) to encode the entirety of the HDR stream in an LDR stream, and may encode meta-data as well. Figure 9.1 gives a general overview of single stream methods. The dual stream methods, for the most part, split the stream into two; a stream which is, usually, backwards compatible, and a residual. Figure 9.2 provides an overview of two stream methods. They will typically use an LDR encoder

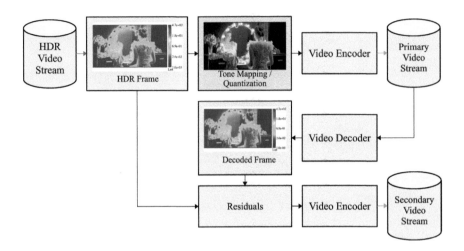

Figure 9.2. The encoding pipeline for dual stream methods. (The original HDR video is courtesy of Jan Fröhlich.)

to encode both streams, and are designed with 8-bit encoders in mind, particularly since most were designed when 8-bit encoders were the standard. Frequently, some meta-data may also be stored.

9.1 Perception-Motivated High Dynamic Range Video Encoding

Mantiuk et al. [245] proposed one of the first solutions for storing HDR videos in an efficient way called HDRV; see Figure 9.3. This is based on the MPEG-4 part 2 standard [275] using Xvid [426] as the starting implementation. Some stages were modified to be HDR compatible.

The first step of the encoding is the conversion of the stream from RGB or XYZ into LogLuv [214] with luminance in the linear domain. Then, the luminance dynamic range is reduced taking into account limits of the HVS. This ensures that the quantization error is always below visibility thresholds of the human eye. The mapping function is defined as a solution of an ordinary differential equation for ψ, the backward mapping function, which is unknown:

$$\frac{d\psi(L_{\mathrm{w}})}{dL_{\mathrm{w}}} = \frac{2}{a}\mathrm{TVI}(\psi(L_{\mathrm{w}})), \tag{9.1}$$

$$\psi(0) = 10^{-4}\mathrm{cd/m}^2, \qquad \psi(L_{\mathrm{w,\,max}}) = 10^8\mathrm{cd/m}^2,$$

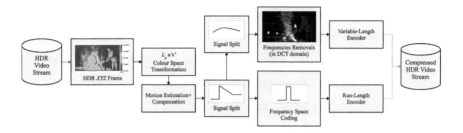

Figure 9.3. The encoding pipeline for HDRV by Mantiuk et al. [245]. (The original HDR video is courtesy of Jan Fröhlich.)

where $l_{max} = 2^n - 1$, TVI is the threshold versus intensity function introduced by Ferwerda et al. [130] (see Section 3.2.4), and $a > 0$ is a parameter that defines how much lower/conservative the maximum quantization errors are compared to TVI. Note that Equation (9.1) assumes local adaptation at the pixel.

After this step, motion estimation and inter-frame prediction is performed as in standard MPEG-4 part 2 (see [275] for more details). Subsequently, non-visible frequencies are removed in the frequency domain using the discrete cosine transform (DCT). As before, this step is not modified, keeping even the same quantization matrices of the standard. However, a correction step is added after frequencies removal to avoid ringing artifacts around sharp transition, for example, an edge between a light source and a diffuse surface. This step separately encodes strong edges into an edge map using run length encoding and the other frequencies into DCT coefficients using variable length encoding.

The decoding of a key frame is straightforward. First, the edge map and DCT coefficients are decoded from the encoded stream. Second, the two signals are recombined. Third, ψ is applied to the luminance channel, obtaining the final world luminance. Finally, the pixel values are converted back from Luv color space into XYZ or RGB color space. When P-frames or B-frames are decoded, an additional step of reconstruction is added using motion vectors. See the MPEG-4 part 2 standard [275] for more details.

9.1.1 Results and Evaluation

The method was tested using different scenes, including rendered synthetic videos, moving HDR panoramas from Spheron Camera [351], and a grayscale Silicon Vision Lars III HDR [394]. The results showed that HDRV can achieve compression rates of around 0.09 bpp–0.53 bpp, which is approximately double the amount of MPEG-4 with tone mapped HDR videos. HDRV outperforms OpenEXR, which reaches rates of around 16–28 bpp.

More recent results [276] (see Section 10.5.1) show that this method performs very well compared to other popular methods in both subjective and objective evaluations.

9.2 Backward Compatible HDR-MPEG

Backward compatible HDR-MPEG is a codec for HDR videos that was introduced by Mantiuk et al. [244]. This algorithm is an extension to the standard MPEG-4 part 2 standard [275] that works on top of the standard encoding/decoding stage allowing backward compatibility. Each frame is divided in an LDR part, via tone mapping, and an HDR part, and a reconstruction function (RF), a numerical inverse TMO, is employed for decoding.

The encoding stage takes as input an HDR video in the XYZ color space and it applies tone mapping to each HDR frame, obtaining LDR frames as the first step. Figure 9.4 shows the complete pipeline. These are coded using Xvid, an MPEG4 part 2 implementation, stored in an LDR stream, and finally decoded to obtain uncompressed and MPEG quantized frames. After this, the LDR frame and the HDR frame are converted to a common color space. For both HDR and LDR frames, CIE 1976 uniform chromaticity scales (u', v') coordinates are used to code chroma. While non-linear luma of $sRGB$ is used for LDR pixels, a different luma coding is used because $sRGB$ non-linearity is not suitable for high luminance ranges $[10^{-5}, 10^{10}]$ (see [244]). This luma coding, at 12 bits, for HDR luminance values is given as

$$l_{\mathrm{w}} = f(L_{\mathrm{w}}) = \begin{cases} 209.16 \log(L_{\mathrm{w}}) - 731.28 & \text{if } L_{\mathrm{w}} \geq 10469, \\ 826.81 L_{\mathrm{w}}^{0.10013} - 884.17 & \text{if } 5.6046 \leq L_{\mathrm{w}} < 10469, \quad (9.2) \\ 17.554 L_{\mathrm{w}} & \text{if } L_{\mathrm{w}} < 5.6046. \end{cases}$$

At this point, the HDR and LDR frames are in a comparable color space. Then RF, which maps LDR values, l_{d}, to HDR ones, l_{w}, is calculated by averaging l_{w} values that fall into one of 256 bins representing the l_{d} values:

$$\mathrm{RF}(i) = \frac{1}{|(\Omega_i)|} \sum_{\mathbf{x} \in \Omega_i} l_{\mathrm{w}}(\mathbf{x}), \qquad \text{where } \Omega_i = \big\{ \mathbf{x} | l_{\mathrm{d}}(\mathbf{x}) = i \big\},$$

where $i \in [0, 255]$ is the index of a bin. RF for chromaticity is approximated by $(u'_{\mathrm{d}}, v'_{\mathrm{d}}) = (u'_{\mathrm{w}}, v'_{\mathrm{w}})$. Once RF functions are calculated for all frames, they are stored in an auxiliary stream using Huffman encoding. After this stage, a residual image is calculated for improving overall quality, especially in small details for luma:

$$r_l(\mathbf{x}) = l_{\mathrm{w}}(\mathbf{x}) - \mathrm{RF}(l_{\mathrm{d}}(\mathbf{x})).$$

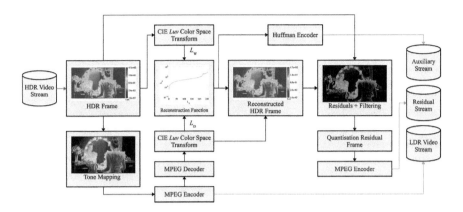

Figure 9.4. The encoding pipeline for backward compatible HDR-MPEG by Mantiuk et al. [244]. (The original HDR video is courtesy of Jan Fröhlich.)

The residual image is discretized at 8 bits, using a different quantization factor for each bin based on its maximum magnitude value. This leads to

$$\hat{r}_l(\mathbf{x}) = \left\lceil \frac{r_l(\mathbf{x})}{q(m)} \right\rceil_{-127}^{127}, \qquad \text{where } m = k \Leftrightarrow i \subset \Omega_k,$$

where $q(m)$ is the quantization factor, which is calculated for a bin Ω_l as

$$q(m) = \max\left(q_{\min}, \frac{\max_{\mathbf{x} \in \Omega_l}(|r_l(\mathbf{x})|)}{127} \right).$$

Then, $\hat{r}_l$ needs to be compressed in a stream using MPEG. A naïve compression would generate a low compression rate, because a large amount of high frequencies are present in $\hat{r}_l$. In order to improve the compression rate, the image is filtered removing frequencies in regions where the HVS cannot perceive any difference. This is achieved using the original HDR frame as guidance when filtering. The filtering is performed in the wavelet domain, and it is applied only to the three finest scales modeling contrast masking and lower sensibility to high frequencies.

When decoding, the MPEG streams (tone mapped video and residuals) and RF streams are decoded. Then, an HDR frame is reconstructed applying firstly its RF to the LDR decoded frame and secondly adding residuals to the expanded LDR frame. Finally, CIE Luv (see Appendix B) values are converted to XYZ ones by inverting Equation (9.2) for luminance.

9.2.1   Results and Evaluation

HDR-MPEG was evaluated using three different metrics: HDR-VDP [243, 246], universal image quality index (UQI) [403], and classic signal-to-noise ratio (SNR). Initially a study that explored the influence of a TMO on quality/bit rate was conducted using the time-dependent visual adaption operator [302], the fast bilateral filtering operator [118], the photographic tone reproduction operator [323], the gradient domain operator [129], and the adaptive logarithmic mapping operator [116]. These TMOs were modified to avoid temporal flickering and applied to a stream using their default parameters. The conclusion of the study showed that most of these TMOs have the same performances except for the gradient domain one, which created larger streams. However, this TMO generated images better suited for backward compatibility. Therefore, the choice of a TMO for the video compression depends on the trade-off between bit rate and the backward compatible quality. The second study compared HDR-MPEG against HDRV [245] and JPEG-HDR [410] using the photographic tone reproduction operator as the TMO [323]. The results showed that HDR-MPEG has a better quality than JPEG-HDR, but a similar one to HDRV.

9.3   Rate-Distortion Optimized Compression

Lee and Kim [220] proposed a backward compatible HDR video compression method based on tone mapping and residuals. The new key contribution is the use of a temporal coherent TMO to avoid flickering and an optimization process to automatically allocate bits to tone mapped frames and residuals.

The first step of the encoding is to tone map each frame of the stream with a temporal coherent version of the gradient domain operator [219]; see Section 5.4.1. Figure 9.5 shows the full pipeline. After tone mapping, the stream is encoded using the H.264 standard [418], and luminance residuals are encoded as

$$R(\mathbf{x}) = \log_2\left(\frac{L_{\mathrm{w}}(\mathbf{x})}{L_{\mathrm{d}}(\mathbf{x})} + \epsilon\right), \qquad \epsilon > 0,$$

where $L_{\mathrm{w}}(\mathbf{x})$ is the HDR luminance at pixel coordinate $\mathbf{x}$ of the original frame I_{w}, and $L_{\mathrm{d}}(\mathbf{x})$ is the decoded luminance of the LDR stream at same pixel coordinate and frame. The calculation of R can lead to noise due to quantization, which is removed by filtering it using the cross bilateral filter with $L_{\mathrm{w}}(\mathbf{x})$ as guidance. Finally, the residual stream is encoded using H.264 as well.

When encoding using H.264, the bit rates of the two streams are not the same, but it is optimized in order to increase compression rates. The

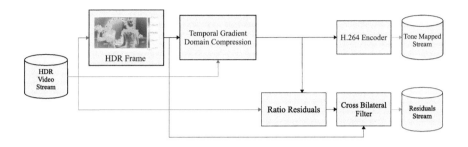

Figure 9.5. The encoding pipeline for rate-distortion optimized compression of HDR videos, Lee and Kim [220]. (The original HDR video is courtesy of Jan Fröhlich.)

quantization parameters for the LDR sequence, QP_d, and the ratio sequence, QP_{ratio}, are calculated such that distortions of the reconstructed LDR frames, D_d, and ones of reconstructed HDR frames, D_w, are minimized. This problem can be defined as a Lagrangian multiplier minimization problem:

$$J = D_w + \mu D_w + \lambda(R_d + R_{ratio}),$$

where μ and λ are two Lagrangian multipliers. The authors found from the analysis of J, a formula for controlling the quality of ratio stream as

$$QP_{ratio} = 0.77 QP_d + 13.42.$$

When decoding, the two H.264 streams are decoded, and the HDR frame is reconstructed as

$$\tilde{I}_{w,i}(\mathbf{x}) = I_{d,i}(\mathbf{x}) \cdot 2^{R(\mathbf{x})} \quad \forall i \in \{R, G, B\},$$

where $\tilde{I}_{w,i}$ is the current reconstructed HDR frame for the i-th color channel, and I_d is the decoded tone mapped LDR frame.

9.3.1   Results and Evaluation

This compression method was evaluated against MPEG-HDR [244]. The metrics used were PSNR for the tone mapped backward compatible frames, and HDR-VDP [246] for the HDR frames. The results showed that while the proposed method has a better quality than MPEG-HDR at low bit rates for HDR frames, on average 10% less HDR-VDP error, MPEG-HDR has better quality at bit rates higher than 1 bpp, on average 2–5%. Regarding tone mapped frames, the rate-distortion optimized method has on average more than 10 dB better quality than MPEG-HDR at any bit rate. Finally, the authors analyzed the bit rates of tone mapped and residual streams and showed that, on average, 10–30% more space is needed for supporting HDR videos.

9.4 Temporally Coherent Luminance-to-Luma Mapping

Garbas and Thoma [139] presented a non-backwards compatible HDR video compression method by making efficient use of n-bit in higher bit depth encoders for a single stream method. This method adopts a similar representation to that of LogLuv (see Section 2.2), however, the authors argue that the amount of detail maintained in LogLuv goes beyond what is perceivable by the HVS. They recommend a more efficient form of encoding that uses a smaller bit depth and adapts to the available content, while maintaining temporal coherence across frames suitable for encoding video.

The method adapts the luminance to luma mapping at n-bit that is based on the following:

$$L_{\mathrm{d}} = \left\lfloor \frac{2^n - 1}{\log_2(L_{\mathrm{w,\ max}}/L_{\mathrm{w,\ min}})} \left(\log_2(L_{\mathrm{w}}) - \log_2(L_{\mathrm{w,\ min}}) \right) \right\rceil, \qquad (9.3)$$

where L_{w} represents the HDR content, L_{d} the converted data at a bit depth of n, and $L_{\mathrm{w,\ min}}$ and $L_{\mathrm{w,\ max}}$ are, respectively, the minimum and maximum HDR values. Similarly the inverse equation is computed as

$$L_{\mathrm{w}} = 2^{(L_{\mathrm{d}}+0.5)\frac{\log_2(L_{\mathrm{w,\ max}}/L_{\mathrm{w,\ min}})}{2^n-1}+\log_2(L_{\mathrm{w,\ min}})}. \qquad (9.4)$$

The dynamic nature of video content entails that $L_{\mathrm{w,\ min}}$ and $L_{\mathrm{w,\ max}}$ would be different for different, possibly contiguous, images within a video stream. Hence, the resultant values L_{d} will not map to the same values which can result in a loss in efficiency from temporal prediction and possible temporal artifacts. For two consecutive frames k and $l = k+1$ the value of $L_{\mathrm{d},l}$ could be computed relative to $L_{\mathrm{d},k}$ via Equation (9.3) and Equation (9.4) as

$$\begin{aligned} L_{\mathrm{d},l} &= \ (L_{\mathrm{d},l} + 0.5)\frac{\log_2(L_{\mathrm{w,\ max,}\ k}/L_{\mathrm{w,\ min,}\ k})}{\log_2(L_{\mathrm{w,\ max,}\ l}/L_{\mathrm{w,\ min,}\ l})} + \\ &\quad (2^n - 1)\frac{\log_2(L_{\mathrm{w,\ min,}\ k}/L_{\mathrm{w,\ min,}\ l})}{\log_2(L_{\mathrm{w,\ max,}\ l}/L_{\mathrm{w,\ min,}\ l})} \\ L_{\mathrm{d},l} &= \ (L_{\mathrm{d},l} + 0.5)w + o, \end{aligned}$$

where w and o can be thought of, respectively, as the scale and the offset and can be computed via the max and min values of consecutive frames. This relationship across the two frames is ideal to be used in conjunction with the weighted prediction used in codecs such as H.264 to maintain temporal predictions for smooth transitions between different sequences. The range of values for w and o are limited and can take on values between -128 and 127. For frames k and l, the maximum $\hat{L}_{w, \max, l}$ and minimum $\hat{L}_{w, \min, l}$, for computed values $\hat{L}_w$, need to be found to meet:

$$\frac{\log_2(\hat{L}_{w, \max, k}/\hat{L}_{w, \min, k})}{\log_2(\hat{L}_{w, \max, l}/\hat{L}_{w, \min, l})} 2^{\log WD} = w; \{w \in \mathbb{Z} | -128 \le w \le 127\}, \quad (9.5)$$

$$\frac{\log_2(\hat{L}_{w, \min, k}/\hat{L}_{w, \min, l})}{\log_2(\hat{L}_{w, \max, l}/\hat{L}_{w, \min, l})} \frac{2^n - 1}{2^{n-8}} = o; \{o \in \mathbb{Z} | -128 \le w \le 127\}, \quad (9.6)$$

under $\hat{L}_{w, \max, l} \ge L_{w, \max, l}$ and $\hat{L}_{w, \min, l} \le L_{w, \min, l}$. $2^{\log WD}$ and 2^{n-8} are the quantization levels to represent w and o, respectively. The initial values are set to $\hat{L}_{w, \max, l} = L_{w, \max, l}$ and $\hat{L}_{w, \min, l} = L_{w, \min, l}$ and solving Equation (9.5) and Equation (9.6) for $\hat{L}_{w, \max, l}$ and $\hat{L}_{w, \min, l}$ as

$$\hat{L}_{w, \min, l} = 2^{[\log_2(L_{w, \min, k}) - \frac{o \cdot 2^{\log WD} 2^{n-8}}{w(2^n - 1)} \log_2(L_{w, \max, k}/L_{w, \min, k})]}$$

$$\hat{L}_{w, \max, l} = 2^{[\frac{2^{\log WD}}{w} \log_2(L_{w, \max, k}/L_{w, \min, k}) + \log_2(\hat{L}_{w, \min, l})]}.$$

If conditions from Equation (9.5) and Equation (9.6) are violated, w is decreased or o is increased and the equations are recomputed. Once $\hat{L}_{w, \min, l}$ and $\hat{L}_{w, \max, l}$ are computed, they are input into the original Equation (9.3).

The method also takes into account the varying quantization of the luminance space and identifies a quantization parameter (QP) that can be adapted each frame based on the relation between consecutive frames as follows:

$$Q_{\mathrm{rel}, l, k} = \frac{Q_{\mathrm{step}, l}}{Q_{\mathrm{step}, k}} = \frac{\log_2(L_{w, \max, k}/L_{w, \min, k})}{\log_2(L_{w, \max, l}/L_{w, \min, l})},$$

for quantization steps Q_{step}. Since steps in QP values approximately double every six units increase, the change in QP ΔQP is given as

$$\Delta QP_{l,k} = \mathrm{round}\left(6\log_2(Q_{\mathrm{rel}, l, k})\right),$$

and is used by initially adjusting subsequent fame QPs by that of the original frame.

9.4.1   Implementation and Results

The method converts from RGB to LogLuv and then encodes the results using a high bit-depth encoder. The authors report using 12 bits for the luma component and 8 bits for both the u and v channels; u and v are also sub-sampled by a factor of two. However, the method could be adapted to other bit depths. Results are presented for three scenes and compared across bit rates versus a non-adaptive method and a method that is not temporally coherent. The proposed method performs well compared to the other two, demonstrating both the adaptive and temporally coherent nature provide a good performance gain in terms of quality per bit rate. Results by Mukharjee et al. [276] (see Section 10.5.1) show that this method performs well compared to other HDR compression methods.

9.5   Optimizing a Tone Mapping Curve for Encoding

The majority of backwards compatible methods make use of tone mapping methods in order to provide a format that is compatible with existing encoding/decoding ecosystems. This typically takes the form of either the tone mappers considered best or the one chosen/preferred by the user. Mai et al. [237] provide an alternative solution. Noting that the tone mapping operation has a considerable influence on the quality of the displayed content once decoded, the proposed method attempts to identify the tone mapping curve that could produce the decoded HDR content which is closest to the original. This is conducted via a process of optimization at the tone mapping stage. The solution replaces user decision in the choice of tone mapper by a tone curve which is generated specifically based on the requirements of the encoding and decoding process.

The original content L_w is initially tone mapped to L_d. L_d is subsequently encoded using a preferred LDR encoder and subsequently decoded to $\widetilde{L}_\text{d}$. The original tone mapping curve is inverted and used to reconstruct the HDR $\widetilde{L}_\text{w}$. Ideally the error $|\widetilde{L}_\text{w}\text{-}L_\text{w}|$ is as small as possible. In the proposed solution the tone mapping curve is chosen such that this error is minimized.

The proposed solution represents the tone mapping curve as a piecewise linear function composed of monotonically increasing nodes $(l_k,\, d_k)$ such that the distance between two successive nodes $(l_k,\, d_k)$ to $(l_{k+1},\, d_{k+1})$ consists of a fixed luminance δ; see Figure 9.6. The tone-mapping curve consists of a sequence of gradients:

$$s_k = \frac{d_{k+1} - d_k}{\delta},$$

and the tone mapping curve is thus:

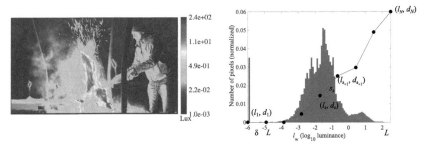

Figure 9.6. An example of the tone mapping curve proposed by Mai et al. [237]: (a) The input HDR frame (false color) from an HDR video. (b) The histogram of the frame in (a) with the tone mapping curve over-imposed on it. (The original HDR video is courtesy of Jan Fröhlich.)

$$L_\mathrm{d} = (l_\mathrm{w} - l_k)s_k + d_k; \qquad l_\mathrm{w} = \log_{10} L_\mathrm{w},$$

for $l_k \leq l \leq l_{k+1}$. The decoding function converts L_d to $\widetilde{L_\mathrm{d}}$. $\widetilde{L_\mathrm{w}}$ is then computed from $\widetilde{L_d}$ and s_k as

$$\log_{10} \widetilde{L_\mathrm{w}}(L_\mathrm{d}, s_k) = \widetilde{l}_\mathrm{w}(L_\mathrm{d}, s_k) = \begin{cases} \frac{L_\mathrm{d} - d_k}{s_k} + l_k & \text{if } s_k > 0, \\ \sum_{l \epsilon s_0} l \cdot P_L(l) & \text{if } s_k = 0, \end{cases} \tag{9.7}$$

for $s_k \epsilon \{s_1, \ldots, s_N\}$. $P_L(l)$ is the probability of an HDR pixel having that value and is computed from the histogram of the image being compressed. For $s_k = 0$, $P_L(l)$ is used to estimate $\widetilde{L}_\mathrm{w}$.

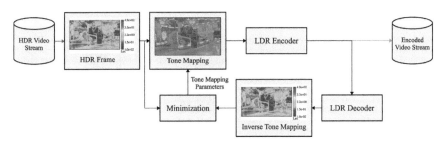

Figure 9.7. The full optimization of tone curve pipeline. Note this method is computationally expensive, hence, an alternative practical proposition as seen in Figure 9.8 is proposed. (The original HDR video is courtesy of Jan Fröhlich.)

Figure 9.7 demonstrates the overall process. Ideally, the method would be applied to optimizing the tone curve based on an iterative process which modifies the tone curve parameters, encodes the tone mapped image, decodes it and uses the inverse of the tone curve to generate the decoded HDR content as described above. The mean square error could be computed as $\|L_w - \widetilde{L_w}\|_2^2$ and the s_k parameters could be chosen such as to minimize this error. However, following this entire process would be too computationally expensive to conduct per frame, or sequence of frames, hence the method presents a solution based on a statistical distortion model and a closed form solution. Figure 9.8 demonstrates the more practical alternative to Figure 9.7.

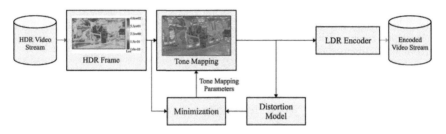

Figure 9.8. The pipeline proposed for improving performance when optimizing a tone curve for HDR encoding. (The original HDR video is courtesy of Jan Fröhlich.)

9.5.1 Improving Computational Efficiency

The method presents a statistical distortion model in order to predict the error generated from the coding and decoding process and uses it to estimate the error as follows:

$$E\left[\|L_w - \widetilde{L_w}\|_2^2\right] = \sum_{l_{w,\min}}^{l_{w,\max}} \sum_{L_d=0}^{L_{d,\max}} (\widetilde{l_w}(L_d, s_k) - l_w)^2 P_c(L_d - \widetilde{L_d}|L_d) \cdot P_L(l_w),$$

$$(9.8)$$

where $P_c(L_d - \widetilde{L_d}|L_d)$ is the probability that the error from encoding is $L_d - \widetilde{L_d}$, and $P_L(l_w)$ is computed from the histogram as described above. The expected error depends primarily on the values of the tone mapping curve defined by s_k, hence the expected error value $E[\|L_w - \widetilde{L_w}\|_2^2]$ can be viewed as a function $\varepsilon(s_k)$. Furthermore, the compression probability is independent of L_d Equation (9.8) can then be written as

$$\varepsilon(s_k) = \sum_{l_{w,\min}}^{l_{w,\max}} \sum_{L_d=0}^{L_{d,\max}} (\widetilde{l_w} - l_w)^2 P_c(L_d - \widetilde{L_d}) \cdot P_L(l_w). \qquad (9.9)$$

The authors state that the probability $P_c(L_d - \widetilde{L_d})$ can be estimated for any scheme and they provide a method for H.264 I-frame coding. Once these values are computed, the optimization can be stated as a minimization of the slopes s_k as follows:

$$\arg\min_{s_1 \ldots s_N} \varepsilon(s_k),$$

for a monotonically increasing function and $s_{\min} \leq s_k \leq s_{\max}$ for $k = 1 \ldots N$ and $\sum_1^N s_k \delta = L_{d, \max}$. $s_{\min}$ is set to a very low value and $s_{\max} = (\log_{10}(1.01))^{-1}$ complying with the luminance detection threshold of 1%.

The above statistical model requires the computation of $P_c(L_d - \widetilde{L_d})$ and significant iterations to solve. The authors present a subsequent model that provides a closed-form solution. By making the assumption that the slopes at pixel values L_d and $\widetilde{L_d}$ are the same and by substituting Equation (9.7) in Equation (9.9) the following can be obtained:

$$\varepsilon(s_k) = \sum_{l_{w, \min}}^{l_{w, \max}} \sum_{L_d=0}^{L_{d, \max}} \left(\frac{L_d - \widetilde{L_d}}{s_k} \right)^2 P_c(L_d - \widetilde{L_d}) \cdot P_L(l_w),$$

which yields:

$$\varepsilon(s_k) = \sum_{l_{w, \min}}^{l_{w, \max}} \frac{P_L(l_w)}{s_k^2} Var(L_d - \widetilde{L_d}). \tag{9.10}$$

The advantage of this solution is the lack of dependence on the compression probability. The optimization can then be expressed as

$$\arg\min_{s_1 \ldots s_N} \sum_{k=1}^N \frac{p_k}{s_k^2}, \quad \text{subject to} \quad \sum_{k=1}^N s_k = \frac{L_{d, \max}}{\delta}, \tag{9.11}$$

where $p_k = \sum_{l=l_k}^{l_{k+1}} P_L(l)$ and can be solved analytically to obtain:

$$s_k = \frac{L_{d, \max} \cdot p_k^{\frac{1}{3}}}{\delta \sum_{k=1}^N p_k^{\frac{1}{3}}}. \tag{9.12}$$

9.5.2 Implementation and Results

The method is run by either the statistical distortion model or the closed form solution. The authors report that the closed-form solution is significantly faster that the statistical model. The s_k parameters need to be stored with the content in order to be able to reconstruct the HDR image. The method is computed in $\log_{10}$ luminance space to account for the nonlinear response to luminance of the human visual system. When catering for video the authors recommend the use of a low pass filter over the tone mapping curves to avoid flickering issues that may arise.

Results of the statistical distortion model and closed form model are compared to a fully fledged ground truth optimization and are found to be very competitive in spite of the substantially decreased computational cost. The authors also provide results for the comparison of the proposed method with three TMOs, the photographic tone reproduction operator [323], adaptive logarithmic TMO [116], and the display adaptive TMO [242]. Results were averaged over 40 images. Both MSE and SIIM results showed an improvement in performance by both the statistical distortion method and the closed-form solution compared to the other TMOs.

9.6   Perceptual Quantizer

The Perceptual Quantizer (PQ) [270] is one of the two transfer functions contained in the ITU-R Recommendation BT.2100 [384] specifically on HDR, with the other being Hybrid Log Gamma discussed below. It is also the main transfer function for HDR10 whereby it is quantized to 10-bit which is, at the time of writing, the main format being used in commercial HDR televisions and display devices. PQ is is based on research in vision science, primarily the CSFs at different luminance composed by Barten [55]. It is designed primarily for operation with displays such as television and cinema and thus content is display referred with a current upper limit of 10,000 cd/m^2.

PQ is relatively straightforward, the encoding method initially linearly normalizes the original HDR content L_w to L_n such that $L_n \in [0, 1]$ the content before applying Equation (9.13).

$$L_d(\mathbf{x}) = \left(\frac{L_n(\mathbf{x})^{\frac{1}{m}} - c_1}{c_2 - c_3 L_n(\mathbf{x})^{\frac{1}{m}}} \right)^{\frac{1}{n}}, \tag{9.13}$$

where L_d is the resultant display adapted value, L_n is the input normalized content, c_1, c_2, c_3, n, and m are all constants. $c_1 = 0.8359$, $c_2 = 18.8516$, $c_3 = 18.6875$, $m = 78.8438$, and $n = 0.1593$.

The function is applied directly on each of the RGB channels or can be applied to different color spaces such as LMS or IPT; see Appendix B. The resultant is then quantized to 10 or 12-bit before being encoded by a traditional LDR encoder using the higher bit depth encoding functionality available in newer encoders.

When decoding the function is inverted as follows:

$$L_n(\mathbf{x}) = \left(\frac{L_d(\mathbf{x})^{\frac{1}{n}} - c_1}{c_2 - c_3 L^{\frac{1}{n}}} \right)^{\frac{1}{m}},$$

where L_n, L_d, c_1, c_2, c_3, m and n are as above. Subsequently L_n is denormalized to be between 0 and 10,000 or adjusted according to the display parameters. Note that both the encoding and decoding curves can be stored as look up tables depending on hardware restrictions in the encoding and decoding platforms.

During evaluation, the authors ran a series of visual tests on a Dolby PRM-4200 HDR monitor by generating different images and encoding them using different bit-depths from 7 bits to 12 bits. The results show that the method does not show visible banding artifcats, and the ideal number of bits for encoding is between 10 and 11 bits.

9.7   Hybrid Log Gamma

Hybrid Log Gamma (HLG) [73] is the other transfer function that forms part of the HDR focused ITU-R Recommendation BT.2100 together with PQ. It follows a similar design philosophy of transferring the incoming HDR content via a curve and using the modified content for encoding with the use of traditional encoders and inverting the process when decoding for displays. As with PQ there is no suggestion of the need to store residual content. The encoding stage is based on a transfer function which attempts to obey a gamma-like curve at lower lighting levels, also following the De Vries–Rose law as PQ does. For brighter areas a logarithmic approach is adopted which matches the HVS's predisposition to luminance. The transfer function adopted is:

$$L_d(\mathbf{x}) = \begin{cases} r\sqrt{L_w(\mathbf{x})} & 0 \leq L_w(\mathbf{x}) \leq 1, \\ a \ln(L_w(\mathbf{x}) - b) + c, & \text{otherwise,} \end{cases} \qquad (9.14)$$

where $a = 0.17883277$, $b = 0.28466892$ and $c = 0.55991073$. r corresponds to the reference white level set to 0.5. This has the effect of generating a "knee" like response and at the lower brightness levels can be used for LDR displays without changes as the expected curve observes gamma relatively well. The encoding method used by the curve makes this method display independent as it does not require an understanding of the display characteristics to store the content.

The method is decoded for displaying via the inverse of the transfer function in Equation (9.14) as

$$L_{\mathrm{w}}(\mathbf{x}) = \begin{cases} \left(\frac{L_{\mathrm{d}}(\mathbf{x})}{r}\right)^2 & 0 \leq L_{\mathrm{d}}(\mathbf{x}) \leq r, \\ \exp\left(\frac{L_{\mathrm{d}}(\mathbf{x}) - c}{a}\right) + b & \text{otherwise.} \end{cases}$$

The evaluation process, which was a subjective study with 33 participants, had the main goal of determining the method capability in order to avoid banding artifacts when content is displayed on an HDR display (tested on the Dolby PRM 4220 monitor). The results of this experiment show that the HLG encoding method is less than or about one grade of impairment, i.e., banding artifacts are *imperceptible* or *just perceptible but not annoying* in the worst case.

9.8   Others Methods

The increase in popularity in HDR content has led to a large number of other HDR video compression methods that have been recently being developed recently. A brief overview of these is given.

9.8.1   Base and Detail Methods

Banterle et al. [37] described a method for HDR video compression based on a dual stream approach. Both streams are designed to be compatible with standard LDR encoders. The first stream, termed base stream, is composed of the luminance component and the second stream, the detail, contains the chromatic residual from the original content.

When encoding, the base stream is calculated by computing the luminance from the original HDR frame, I. Then, the output is passed through an edge-preserving bilateral filter (see Appendix A) and this luminance channel is tone mapped. The tone mapped version is quantized and ready for encoding. It is subsequently decoded, and inverse tone mapped obtaining $\tilde{B}_{\mathrm{w}}$. At this point, detail frame, D, is computed as

$$D_i(\mathbf{x}) = \frac{I_{\mathrm{w},i}(\mathbf{x})}{\tilde{B}_{\mathrm{w}}(\mathbf{x})}, \quad \forall i \in \{R, G, B\},$$

which is subsequently quantized and encoded obtaining $\tilde{D}$. The decoding approach de-quantizes both the base and detail layers and, subsequently, the base layer is inverse tone mapped. The HDR content is then retrieved as $\tilde{I}_{\mathrm{w},k}(\mathbf{x}) = \tilde{B}_{\mathrm{w}} \times D_i(\mathbf{x})$, for $i \in \{R, G, B\}$. This method requires two streams, although one of the streams could be a single channel stream. This method is not considered backwards compatible as neither of the streams contains data designed to be viewed using legacy systems. Results for this method, reported by Mukharjee et al. [276], using a generic sigmoid for the tone mapper, show that it performs competitively with other two stream methods.

Debattista et al. [105] presented a similar method termed *practical method*. The base frame is computed as above and provides the luminance from the original content. The detail frame is computed as $D_i(\mathbf{x}) = I_{w,i}(\mathbf{x})/(\tilde{B}_w(\mathbf{x}) + 1)$, for $i \in \{R, G, B\}$ allowing D_i to be backwards compatible, essentially combining the tone mapping stage (removing choice of tone mapper) and computation of detail. Decoding is conducted via inverting this function. The evaluation of the method demonstrate that it can achieve competitive quality performance against other two stream methods. However, it does not have the flexibility of other methods that can make use of different TMOs to cope with different artistic needs.

9.8.2 Optimal Exposure Compression

Debattista et al. [106] presented a method based on finding the optimal exposure to fit within the bit-depth of a given stream. This is distinct from other approaches presented as there is no use of a tone mapping or transfer functions. This view is motivated by previous studies showing that there were no significant differences between single exposure content and tone mapped content in subjective comparisons [17], and also by the complexity of parameters and multiplicity of tone mappers available. The data within the optimal range is stored as a stream, which is backwards compatible and at a chosen bit-depth depending on the codec used. The remaining data is encoded in the logarithmic domain and stored as a residual. The optimal method attempts to maximize luminance over a contiguous area for an HDR image, I_w, at exposure, e, where $f(\cdot)$ is defined as

$$f(I_w, e) = \sum_{p \in \text{pixels}} \begin{cases} 1 & \text{if } (2^n - 1) \times \big(I_w(\mathbf{x}_p) \cdot e\big) \in [a, b], \\ 0 & \text{otherwise,} \end{cases}$$

for bit depth n, for each pixel in the image p and if the current exposure scaled by bit depth is within an accepted range $[a, b]$. This is computed via a histogram of luminance in log space and by finding the area of contiguous bins with the largest values; the use of histograms is similar to the automatic exposure method [285]. The number of bins is set automatically via the Freedman–Diaconis rule, which helps generate the optimal exposure frame used for the backwards compatible stream. The residual is computed by subtracting the reconstructed HDR (obtained by the optimal exposure frame) from the original HDR content. Results demonstrate the method is competitive versus other backwards-compatible methods although offering an alternative mode of backwards compatibility.

9.8.3 An HDR Video Codec Optimized by Large-Scale Testing

Eilertsen et al. [122] proposed a method based on the results of large-scale testing of a number of methods over 33 HDR video sequences. Three aspects of the encoding decoding were tested: a high bit-depth encoder, a transfer function for luminance encoding and color encoding. For the encoder, the open source VP9 was chosen due to its quality, bit depth compression up to the required 12 bits, and open source availability. The transfer function was chosen from four tested. The first was a scaled and normalized content in the logarithmic domain, the second based on HDRV (see Section 9.1), the third an extended version based on a CSF, and the final based on the Barten curve similar to the PQ method discussed in Section 9.6. A large difference was noted between the first method and the other three but not that much variation among those three; however, finally the Barten inspired model was chosen due to its more uniform distribution of luminance. For color channels, LogLuv outperformed both RGB and YC_bC_r. The resultant choices inspired the open source Luma HDRv format [1] with VP9 as based codec, the PQ equation at 11-bit for transfer function, and LogLuv color space.

9.8.4 Power Transfer Functions

Hatchett et al. [165] investigated the use of power transfer functions for single stream encoding and decoding HDR video. Such power functions find use in a multitude or applications, particularly the computation of gamma for traditional displays. The use of power transfer functions is suitable due to their relative straightforward and practical nature and their computational efficiency. The encoding is given by $\text{PTF}_\gamma(V) = V^{\frac{1}{\gamma}}$ and the decoding by $\text{PTF}'_\gamma(V) = V^\gamma$ such that encoding an image by a given power value γ is given as $L_d = \text{PTF}_\gamma(L_w)$ and similarly for decoding $\tilde{L}_w = \text{PTF}'_\gamma(\tilde{L}_d)$. A number of values of γ were investigated across a range of content and the best was discovered to be around a value of 4. Therefore, PTF_4 was chosen for further comparative results as it gave the best balance between quality and computational performance, since the decoding only requires two floating point multiplications. The evaluation of the method showed that it has comparable results with other methods in terms of quality. However, it has a superior, even to look up tables, computational performance both in encoding and decoding. This high computational efficiency makes the method suitable for real-time implementations on low-power environments such a set-top boxes.

---

[1] http://lumahdrv.org/

9.9 Summary

Captured real-world lighting results in very large data sizes. Uncompressed, a single HDR pixel requires 12 bytes of memory to store the three single precision floating-point numbers for the RGB values. A single HDR frame of uncompressed 4K UHD resolution (3840×2160 pixels) requires approximately 94.92 Mbytes of storage, and a minute of data at 30 frames per second needs 166 Gbytes, i.e., a lot of data. Without compression, it is simply not going to be feasible to store this amount of data, let alone transmit it. Although terrestrial broadband is regularly increasing in bandwidth, even this is unlikely to be sufficient to broadcast any uncompressed HDR films in the future.

Not only does the compression have to significantly reduce the amount of data that needs storing and transmitting, but it should do so with minimal loss of visual quality. This is a major challenge as the increased luminance means that many artifacts, which may not be noticed in LDR content, will be easily discernable in HDR.

Table 9.1 summarizes HDR video compression methods. An important question that has yet to be resolved is whether new HDR compression algorithms really do need to be backward compatible with existing LDR versions. Ensuring this backward compatibility has the danger of severely limiting the potential novelty and performance of new HDR algorithms and it is clear from recent results that dedicated single stream methods perform better [276]. As the work presented in this chapter has shown, HDR compression is a challenging problem. Much more research is now needed if HDR video are ever to become "the norm" in applications, such as television and video games, and especially for mobile devices.

Method	St	BC	BD	Comments
HDRV [245]	1	no	11	Best performing method in HDR video encoder comparison [276]
Temporal [139]	1	no	14	Performs well in comparison [276]
PQ [270]	1	no	10/12	transfer function for ITU-R Recommendation BT.2100
HLG [73]	1	no	10/12	transfer function for ITU-R Recommendation BT.2100
Luma HDRv [122]	1	no	11	transfer function similar to PQ
PTF [165]	1	no	10/12	High decoding performance
HDR-MPEG [244]	2	yes	8	
Rate Distortion [220]	2	yes	8	Optimizes backwards compatible and residual layer for encoding
Base/detail [37]	2	no	8	
Optimal TC [237]	2	yes	8	Optimizes tone curve
Optimal Exp [106]	2	yes	8/10	Does not use tone mapping

Table 9.1. Summary of the various HDR video compression methods. **St** refers to single or dual stream but does not include any information about any metadata used. **BC** refers to backwards computability with legacy LDR encoders in which one stream will output a viewable LDR stream. **BD** refer to the recommended bit depth based on the application being targeted although this can change in most cases.

As has been shown in previous chapters, many techniques have been proposed for various stages in the HDR pipeline. With such a large number of techniques, it is useful to understand the relative merits of each. As a consequence, several methodologies have now been put forward that evaluate and compare the variety of approaches. Evaluation methodologies provide a better understanding of the relationship between a technique and the image attributes. This can help in the development of the methods more suited to a particular application. Evaluation methodologies may be classified as:

- Metrics. Metrics performed via computational calculations that are used to compare images and videos. These can be straightforward computational differences or may simulate the behavior of the HVS to find perceptual differences between the images/frames.

- Subjective experiments. In this case, experiments with human participants are performed to visually compare stimuli. This typically involves comparing the output results of various techniques with each other and/or with a reference.

10.1 Metrics

Metrics are used to evaluate the similarities between images. They may use different approaches depending on what needs to be achieved. If only objective values need to be compared statistics-based metrics are suitable. If the goal is to understand how two images are perceptually similar, then a simulation of the HVS mechanisms may help to identify perceived differences or similarities between the compared images.

10.1.1   Objective Metrics

A number of image comparison metrics have been proposed in the literature that are based mostly on images statistics. Many of these are not specific for HDR methods but can be applied nonetheless. However, changes to the methods to be suitable for HDR are frequently adopted as discussed below.

Metrics based on Mean Square Error   The mean square error (MSE) between two signals X and Y with N samples is typically computed as

$$\text{MSE}(X,Y) = \frac{1}{N} \sum_{i=1}^{N} \Big(X(i) - Y(i) \Big)^2. \tag{10.1}$$

Listing 10.1 shows the MATLAB code for the computation of MSE between two images based on Equation (10.1). This code along with others in the section are found under the `Metrics` folder.

```
function mse = MSE(img_ref , img_dist)
%check if images have same size and type
[img_ref , img_dist] = checkDomains(img_ref , img_dist);
%compute squared differences
delta_sq = (img_ref - img_dist).^2;
%compute MSE
mse = mean(delta_sq(:));
end
```

Listing 10.1. MATLAB code: MSE.

Another commonly used method RMSE is defined as

$$\text{RMSE}(X,Y) = \sqrt{\text{MSE}(X,Y)}. \tag{10.2}$$

Listing 10.2 shows the MATLAB code for the RMSE calculation based on Equation (10.2).

```
function rmse = RMSE(img_ref , img_dist)
%compute the root square of MSE
rmse = sqrt(MSE(img_ref , img_dist));
end
```

Listing 10.2. MATLAB code: RMSE.

Peak signal to noise ratio (PSNR) is another widely used metric, which takes into account the maximum value of the signal, and can be defined based on MSE as

$$\text{PSNR}(X,Y) = 20 \log_{10} \left(\frac{D_{\max}}{\text{RMSE}(X,Y)} \right),$$

where D_{max} is the maximum allowable signal intensity; for example with LDR images this would be 255. In general, it is set to the bit depth (BD) representation $2^{BD} - 1$. However, for floating point representations it is common to use the value of the reference image, the larger value of the two signals or the maximum possible value of the display on which the images would be viewed [276].

Listing 10.3 shows the MATLAB code for PSNR. To account for different possibilities of the use of maximum this can be passed in as a parameter `max_value`, otherwise the call to `checkDomains.m` sets `max_value` to the maximum value of the reference image `img_ref`.

```
function psnr = PSNR(img_ref , img_dist , max_value)
%check if images have same size and type
[img_ref , img_dist , ~, mxt] = checkDomains(img_ref , img_dist);

%determine the maximum value
if(~exist('max_value', 'var'))
    max_value = -1000;
end

if(max_value < 0.0)
    max_value = mxt;
end

%compute MSE
mse = MSE(img_ref , img_dist);

if(mse > 0.0)
    %compute PSNR
    psnr = 20 * log10(max_value / sqrt(mse));
else
    disp('PSNR:␣the␣images␣are␣the␣same!');
    psnr = 1000;
end
end
```

Listing 10.3. MATLAB code: PSNR.

When dealing with color images the computation of MSE takes these into account in a number of ways. The difference per pixel per channel is computed and added and averaged over the three channels as

$$\mathrm{MSE}(X,Y) = \frac{1}{3N} \sum_{i,j} \sum_{c \in \{R,G,B\}} \left(X_c(i,j) - Y_c(i,j) \right)^2 . \qquad (10.3)$$

Alternatively, the image can be converted into an alternative color space such as CIE Luv (see Appendix B) and the channel PSNRs can be computed and reported individually. This computation of RMSE and PSNR for color channels is computed in the same fashion by modifying the MSE calculation as in Equation (10.3) or by converting to alternative color spaces and computing channels accordingly.

For computing RMSE and PSNR of HDR images, the methods can be applied directly; however, they do not always produce suitable outcomes. One solution to adapt PSNR to HDR was mPSNR [277] which computes MSE individually for a number of exposures, averages these, and computes the PSNR from the resultant. Other methods include applying logarithmic encoding [425] and Perceptually Uniform (PU) encoding [30] to the signals before passing them to the RMSE and PSNR metrics to take into account the non-linear nature of the HVS's response to light. PU encoding was designed particularly for adapting quality metrics for HDR. It is formulated on a contrast sensitivity function that can predict the HVS's sensitivity to lighting at different luminance values and is described in the form of a look up table. PU-PSNR and PU-SSIM (see below) have been shown to correlate very well with subjective results for HDR content whereas PSNR and RMSE do not [276].

Structural similarity-based image quality assessment. Structural similarity-based image quality assessment (SSIM) [404] is a popular metric which attempts to compare luminance, contrast and structure between two images. It is computed via three individual calculations: l for luminance, c for contrast and s for structure. These are computed as

$$l(x,y) = \frac{2\mu_x\mu_y + C_1}{\mu_x^2 + \mu_y^2 + C_1},$$

$$c(x,y) = \frac{2\sigma_x\sigma_y + C_2}{\sigma_x^2 + \sigma_y^2 + C_2},$$

$$s(x,y) = \frac{\sigma_{xy} + C_3}{\sigma_x^2\sigma_y^2 + C_3},$$

where μ_x and μ_y are the mean of signals x and y, σ_x and σ_y are their standard deviations, and σ_{xy} is the covariance. $C_1 = (K_1 L)^2$ for dynamic range L, computed as $2^{BD} - 1$ for bit depth BD and small constant K_1. $C_2 = (K_2 L)^2$ is computed and similarly. C_3 is computed as $C_3 = C_2/2$. Note that x and y are typically, although not necessarily, selected parts of the full signal X and Y.

The SSIM is then computed as a weighted combination of l, c, and s:

$$\text{SSIM}(x,y) = l(x,y)^\alpha \cdot c(x,y)^\beta \cdot s(x,y)^\gamma,$$

where α, β and γ are values used to change the weighting of the individual components. In the special case of $\alpha = \beta = \gamma = 1$, SSIM can be computed as

$$\mathrm{SSIM}(x, y) = \frac{(2\mu_x\mu_y + C_1)(2\sigma_{xy} + C_2)}{(\mu_x^2 + \mu_y^2 + C_1)(\sigma_x^2 + \sigma_y^2 + C_2)}.$$

The method is typically used locally and the SSIM is applied to areas around a pixel for each pixel and can also be weighted thus adjusting the means, standard deviations and co-variance accordingly; the original article [404] suggested an 11×11 Gaussian kernel for weighting. An individual score per image is given by an average across the computed local segments:

$$\mathrm{SSIM}(X, Y) = \frac{1}{M} \sum_{k=1}^{M} \mathrm{SSIM}(x_k, y_k),$$

where x_k and y_k are, respectively, the local segments at the k-th location, and M is the number of local windows of the image.

As with the metrics above, logarithmic and PU encoding [30] can also be used with SSIM. SSIM is available in MATLAB as part of the Image Processing Toolbox (`ssim.m`), hence no further implementation is provided here.

10.1.2 Predicting Visible Difference in HDR Images

HDR-VDP, the visual difference metric proposed by Mantiuk et al. [243, 246, 248], is designed to model aspects of the HVS by predicting perceptually visible differences in HDR images. The metric is an extension of the existing visual difference predictor (VDP) [100] to HDR imaging. VDP is a very popular metric for LDR images based on a model of the HVS. The flow chart for HDR-VDP is shown in Figure 10.1.

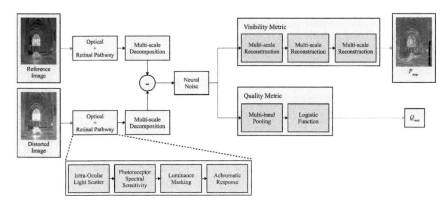

Figure 10.1. Flow chart of the HDR-VDP 2.0 metric by Mantiuk et al. [248].

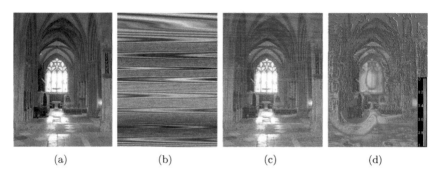

(a)   (b)   (c)   (d)

Figure 10.2. An example of HDR-VDP 2: (a) The original HDR image at f-stop 0. (b) A distortion pattern. (c) Image in (a) added to the distortion pattern (b) at f-stop 0. (d) The result of HDR-VDP 2: blue areas have no perceptual error, green areas have medium error, and red areas have medium-high error. Note that single exposures images are visualized in (a) and (c) to show the differences with the added pattern (b).

As with other metrics, HDR-VDP takes as input the reference and the tested images. This metric generates a probability map where each value represents how strongly the difference between the two images may be perceived by the HVS; see Figure 10.2. The original version of the metric [243, 246] mainly simulates the contrast reduction in the HVS through the simulation of light scattering in the cornea, lens, and retina (optical transfer function [OTF]). This takes into account the nonlinear response of our photoreceptors to light (just noticeable difference [JND]). Because the HVS is less sensitive to low and high spatial frequencies, the CSF is used to filter the input image. Subsequently, the image is decomposed into spatial and orientational channels and the perceived difference is computed (using cortex transform and visual masking blocks). The phase uncertainty step is responsible for removing the dependence of masking on the phase of the signal, and finally the probabilities of visible differences are summed up for all channels generating the difference probability map.

An enhanced version entitled HDR-VDP-2 [248] revamps the original method via calibration and validation against various vision science data sets. The metric contains models that account for the optical and retinal properties based on intra-ocular light scattering, the spectral sensitivity of photoreceptors, luminance masking and achromatic response. This is followed by multi-scale decomposition with the use of steerable pyramids. Neural noise is also modeled via a neural contrast sensitivity function and contrast masking. A custom CSF was also fit for use in many of the above models. An extension entitled HDR-VDP-2.2 [281] improved the pooling method, via the use of a larger dataset of HDR images, and has become the method of choice when using VDP with HDR images. Figure 10.2 gives an example of the method.

10.1.3 Dynamic Range Independent Image Quality Assessment

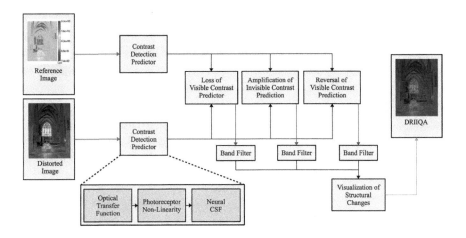

Figure 10.3. Flow chart of the dynamic range independent quality assessment metric of Aydın et al. [31].

Due to the diversity in the dynamic range of images, the available metrics are simply not able to compare images with different dynamic range. For example, it is not possible to compare an HDR image with an LDR image.

Aydın et al. [31] presented a quality assessment metric suitable for comparing images with significantly different dynamic ranges. This metric implements a model of the HVS that is capable of detecting only the visible contrast changes. This information is then used to analyze any visible structure changes. The metric is sensitive to three types of structural changes: loss of visibility contrast, amplification of invisible contrast, and reversal of visible contrast. *Loss of visible contrast* is typically generated by a TMO that strongly compresses details, making them invisible in the tone mapped image. *Amplification of invisible contrast* is the opposite of loss of visible contrast. This is typical when artifacts appear in localized areas of the test image that are not visible in the reference image. *Reversal of visible contrast* occurs when the contrast is visible in both the reference and test images but with different polarity [31].

Figure 10.3 shows the metric's flow chart. A contrast detection prediction step that simulates the mechanisms of the HVS is similar to the one implemented in HDR-VDP. Subsequently, using the cortex transform [412],

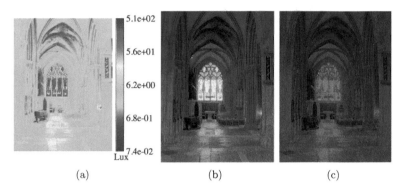

(a)   (b)   (c)

Figure 10.4. An example of the dynamic range independent quality assessment metric example by Aydın et al. [31]: (a) The original HDR image. (b) A tone mapped version of (a). (c) The result of the dynamic range independent quality assessment metric: gray means no perceptual error, green means loss of visibility, blue means amplification of invisible contrast, and red means reversal of visible contrast.

as modified by Daly [100], the output of the contrast detection predictor is subdivided into several bands with different orientations and spatial bandwidths. The conditional probabilities are used to estimate the three types of distortions for each band. Finally, the three types of distortion are visualized with three different colors: green for the loss of visibility contrast, blue for the amplification of invisible contrast, and red for the reversal of visible contrast. Figure 10.4 shows an example of how the independent quality assessment metric appears to the user.

10.1.4   Video Quality Measure for HDR Video

Most HDR metrics are designed for comparing static HDR images and even though they can be used for dynamic aspects they do not take temporal aspects into account. Aydın et al. [33] presented a video quality metric that is dynamic range independent and provides for temporal changes via a temporal model of the HVS. Narwaria et al. [280] presented HDR-VQM, another temporal metric used for comparing video sequences; it has been used in a number of HDR video comparisons recently and is thus expanded upon below.

HDR video quality measure (HDR-VQM) takes into account the HVS's propensity to attend to one given area for a certain amount of time and analyzes changes within a spatio-temporal boundary. The method is based on a number of steps; see Figure 10.5 for its pipeline. These are transformation to emitted luminance, transformation to perceived luminance, error signal computation, and finally, error pooling.

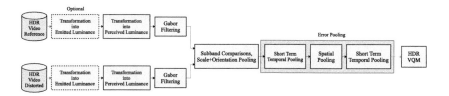

Figure 10.5. HDR-VQM pipeline.

The first phase converts the signal to the range that will be viewed on a given display. This is important with HDR, as different HDR displays have different peak luminance and dynamic range possibly resulting in objectionable artifacts that are visible on some displays but may not be noticeable in others. This is an optional phase as the content may have already been made display referred prior to the use of the metric. The second stage converts the signal into a perceptually-based representation corresponding to the HVS's response to luminance. This is based on PU encoding [30] which maintains a perceptually uniform spacing.

The subband error signal stage generates a spatial error between the reference and distorted signal. log-Gabor filters are employed to find differences at distinct scales and orientations. The subbands are computed via inverse DFT of the product of the signal with the log-Gabor filter. The error is then given as

$$Err_{t,s,o} = \frac{2 l^{src}_{t,s,o} l^{dis}_{t,s,o} + \epsilon}{(l^{src}_{t,s,o})^2 + (l^{dis}_{t,s,o})^2 + \epsilon},$$

where $l^{src}_{t,s,o}$ is the subband value for frame t, scale s and orientation o for the source/reference frame, $l^{dis}_{t,s,o}$ is the corresponding distortion subband and $Err_{t,s,o}$ is the error. ϵ is a small positive constant, which avoids discontinuities. The error per pixel per frame is obtained via equal weighting of scales and orientations:

$$Err_t = \frac{1}{N_{\text{scale}} \times N_{\text{ori}}} \sum_{s=1}^{N_{\text{scale}}} \sum_{s=1}^{N_{\text{ori}}} Err_{t,s,o},$$

where N_{scale} and N_{ori} are, respectively, the number of scales and orientations.

The final stage pools the error in local spatiotemporal neighborhoods and over the entire video sequence. This is computed by dividing the area into three dimensional non-overlapping blocks within local volumes. The standard deviation of each of these volumes provides the error frame ST_{v,t_s} with v being the spatial index and t_s the temporal one. Finally, the metric is computed via both spatial and temporal short-term and long-term pooling as:

$$\text{HDR-VQM} = \frac{1}{|t_s \in L_p|} \times |v \in L_p| \sum_{t_s \in L_p} \sum_{v \in L_p} ST_{v,t_s},$$

where L_p represents the set of the lowest $p\%$ values.

The method was evaluated with a series of 10 HDR video sequences at different qualities ranked by participants. The method outperformed other methods, including HDR-VDP and PuPSNR when participant results were compared with the predicted results from the metric.

10.1.5 Tone Mapped Image Quality Index

Yeganeh and Wang [429] presented a method for directly comparing tone mapped images with the original HDR content. The motivation behind such a metric is to reduce or eliminate the need for time consuming subjective experiments (see Section 10.3) and permits the automation of applications that require error metrics to fine tune and optimize aspects such as parameters and new operators (see Section 4.1.3).

The tone mapped image quality index (TMQI) is made up of two components: one for structural fidelity and the other one for statistical naturalness. The structural fidelity metric takes cues from SSIM, removing the luminance component and maintaining the structure. It is defined as:

$$S_{\text{local}}(x, y) = \frac{2\sigma'_x \sigma'_y + C_1}{\sigma'^2_x + \sigma'^2_y + C_1} \cdot \frac{\sigma_{xy} + C_2}{\sigma_x \sigma_y + C_2},$$

where σ_x and σ_y are the standard deviation for signals x and y, and σ_{xy} is their cross correlation. C_1 and C_2 are stabilizing constants. σ' is a nonlinear mapping to the standard deviation based on the visual sensitivity of contrast. σ' is derived based on psychometric functions and defined as:

$$\sigma' = \frac{1}{\sqrt{2\pi}} \int_{-\infty}^{\sigma} e^{-\frac{(x - t_\sigma)^2}{2\theta_\sigma^2}} \, dx,$$

where $t_\sigma(f) = \overline{\mu}/\sqrt{(2)}\lambda A(f)$ and $\overline{\mu}$ is the mean luminance set to 128 for 8-bit tone mapped LDR images, $A(f)$ is based on a CSF such that $A(f) = 2.6(0.0192 + 0.114f) \exp(-(0.114f)^{1.1})$ and λ is a scaling constant. $\theta_\sigma(f) = t_\sigma(f)/k$ where k is a constant, which is typically set to 3.

As with SSIM, the structural component of TMQI is computed locally and averaged as

$$S_l = \frac{1}{N_l} \sum_{k=1}^{N_l} S_{\text{local}}(x_k, y_k),$$

for N_l patches in the l-th scale. This is then computed at different scales l and pooled to $S = \prod_{l=1}^{L} S_l^{\beta_l}$ for L scales with per-scale weighting vector β_l. The authors recommend an approach similar to that recommended by the original SSIM publication with S_{local} computed at 11×11 Gaussian weighted sliding window. C_1 and C_2 are, respectively, set to 0.01 and 10.

The naturalness component is computed based on measures of brightness and contrast. It is calculated through the use of histograms of mean and standard deviation of 3,000 LDR images for estimates of brightness and contrast. These fit Gaussian and Beta distributions, respectively, as

$$P_m(m) = \frac{1}{\sqrt{2\pi}\sigma_m} e^{-\frac{m-\mu_m}{2\sigma_m^2}} \text{ and } P_d(d) = \frac{(1-d)^{\beta_d-1} d^{\alpha_d-1}}{B(\alpha_d, \beta_d)},$$

where $B(\cdot)$ is the Beta function and the parameters are fitted to the distributions from the aforementioned image database as $\mu_m = 115.94$, $\sigma_m = 27.99$, $\alpha_d = 4.4$, and $\beta_d = 10.1$. The naturalness N is then computed as

$$N = \frac{1}{K} P_m P_d,$$

where $K = \max\{P_m, P_d\}$ to normalize the measure. Finally, TMQI is computed as

$$\text{TMQI} = aS^\alpha + (1-a)N^\beta, \text{ for } 0 \leq a \leq 1.$$

where a controls the importance of S and N, and α and β adjusts the sensitivity. The authors suggest $a = 0.8012$, $\alpha = 0.3046$, and $\beta = 0.7088$ based on fitting the parameters to results of subjective studies.

The method is evaluated across subjective experiments and a strong correlation with subjective scores was found.

10.1.6   A Model for Perceived Dynamic Range

The dynamic range of images is frequently computed as the ratio of the brightest to the darkest pixel, see Section 1.1.1.This objective measure does not necessarily correspond directly with the way humans perceive the dynamic range of an image. Hulusić et al. [176] presented a model for computing the perceived dynamic range of images based on fitting results to captured data. An experiment was conducted with 36 HDR images compared against each other in random groups of 12. In the experiment participants were asked to rate the perceived dynamic range of each image [176, 177]. This was conducted for both chromatic and achromatic images. While the traditional calculation of dynamic range DR (effectively C_R from

Section 1.1.1) correlated well with the captured data, when other metrics were fitted to the data via linear regression a more robust model than DR was found. The model for perceived dynamic range was reported as $0.573DR + 0.448A^{1/4}$ for achromatic images and $0.573DR + 0.448A^{1/4}$ for chromatic images, where DR is as above, and A is the sum of the luminance of all pixels greater than 2,400 cd/m^2 based on findings in the ITU document 6C/146-E [184]. Other metrics such as image key [18] and a measure of local contrast were not found to contribute to the model. This model could predict overall cases with better accuracy than the traditional dynamic range measure but for certain problematic cases, discussed in the original article, where DR struggled this model performed much better.

10.2   Experimental Evaluation Overview

Subjective experimental validation is a traditional way of comparing different methods that serve the same purpose. In HDR this is encountered on a number of occasions. The following will show how subjective experiments can be used for evaluating different tone mapping methods, expansion operators, and compression methods. Furthermore, it can be applied to both static and dynamic stimuli and across different platforms, such as HDR displays, LDR displays and mobile devices. A broad number of methodologies for evaluating different methods exist but they largely conform to a general procedure. Typically, the methods are either compared with each other, and/or with a reference stimulus. Frequently, with HDR content the reference is displayed on an HDR display, while the choice of display for the evaluated methods depends on the type of method tested. For example, tone mappers will most likely be shown on LDR displays while compression methods and expansion operators require the use of an HDR display.

Participants should be chosen with normal or corrected-to-normal vision and carefully instructed about the task that has to be performed. The participant usually performs the task by interacting with a computer interface that presents stimuli and collects data.

There are three broad methodologies that can be applied for comparing stimuli. These are:

- Rating. A participant has to rate an attribute of a stimulus on a scale. A reference or another stimulus can be included. This method can be relatively fast to perform because all other stimuli are not needed for comparison. However, participants can have different perceptions of the rating scale so collected data can be less precise than ranking experiments. Normalization methods are frequently applied to counter this, and to attempt to rectify any gross differences in-between participant results.

- Ranking. Each participant has to rank a series of stimuli based on a criterion. In this case results are more accurate in terms of comparing stimuli as participants need to take a firm decision on the order of presented stimuli. The main disadvantage is that the task is time consuming because a participant has to compare all images; however, it is faster than pairwise comparisons as not all pairs need to be compared individually. Ranking can also be computed indirectly using pairwise comparisons.

- Pairwise comparisons. A participant needs to determine which image in a pair is closer to a reference or which image in a pair better satisfies a certain property. The method produces data with fewer subjective problems, such as rating. Nevertheless, when a full suite of comparisons is performed, the method usually requires more time than rating and ranking. Each participant needs to evaluate $n(t - 1)t/2$ pairs where t is the number of methods compared and n is the number of images. The number of comparisons can be used via certain techniques such as the quicksort method [250].

10.3 Experimental Evaluation of Tone Mapping

Since TMOs dominated most of the early HDR research it is unsurprising that the majority of the first evaluation methods focused on evaluating TMOs. This section provides an overview of this research which generally attempted to evaluate the TMO methods that perform better than others under given circumstances.

One of the first studies on tone mapping evaluation was conducted by Drago et al. [115], where the main goal was to measure the performances of TMOs applied to different scenes based on subjective questions and multidimensional statistical analysis. Kuang et al. conducted a number of studies [208, 209] on the evaluation of TMOs. The results showed that colored and grayscale images were correlated in ranking and that overall preference was correlated with details in dark areas, overall contrast, sharpness, and colorfulness. Moreover, the overall appearance could be predicted by a single attribute. Ledda et al. [216] presented a detailed evaluation of TMOs via pairwise comparisons using an HDR display as a reference; this study is discussed in further detail below. Other studies used real-world scenes for comparisons of TMOs, for example the work of Yoshida et al. [430–432], Ashikhmin and Goyal [28] and Čadík et al. [82,83].

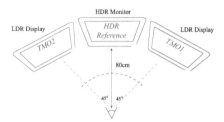

Figure 10.6. An example of the setup for the evaluation of TMOs using an HDR monitor as reference.

Other methods analyzed the use of TMOs for specific conditions, for example, Park and Montag [299] presented two paired-comparison studies to analyze the effect of TMOs on non-pictorial images, referring to images captured outside of the visible wavelength region, such as hyperspectral data and astronomical and medical images.

The above all related to the evaluation of static HDR images, however, recently, due to the increase in number of temporal TMOs, some studies have focused on their evaluation, including the work of Eilertsen el al. [123] and Melo et al. [257, 258]. These are both discussed below. Furthermore, Melo et al. [257, 258] also addressed displaying of TMOs on mobile devices and the challenges faced by small screen devices across different lighting conditions. The use of TMOs on small screen devices was also investigated by Urbano et al. [385] for mobile devices, albeit for still images, and Akyüz et al. [14] for small screen devices represented by a portable camera display.

10.3.1 Paired Comparisons of Tone Mapping Operators Using an HDR Monitor

One of the first studies using an HDR reference and pairwise comparisons was presented by Ledda et al. [216]. In this experiment an HDR monitor was employed to display the ground truth HDR images that the tone mapped images were compared against. The setup of the experiment is shown in Figure 10.6. Forty eight participants took part in the experiments, and 23 images were tone mapped using six TMOs: histogram adjustment operator [215], fast bilateral filtering operator [118], photographic tone reproduction operator [323], iCAM 2002 [125], adaptive logarithmic mapping [116], and local eye adaptation operator [217].

The collected data was analyzed using accumulated pairwise preference scores [103] in combination with coefficients of agreement and consistency [192]. The study showed that, on the whole, the photographic tone reproduction and iCAM 2002 performed better than the other operators, but in the case of grayscale images, the former operator was superior. This is due to the fact the iCAM processes primarily colors, and it is disadvantaged in a grayscale context. Moreover, iCAM 2002 and the local eye adaptation operator performed very well for detail reproduction in bright areas. In conclusion, iCAM 2002 performed generally better than other operators, because colors are an important stimuli in human vision. Furthermore, the study highlighted low performances for the fast bilateral filtering operator. This may be due to the fact that high frequencies are exaggerated with this method when compared to the reference.

10.3.2   Image Attributes and Quality for Evaluation of Tone Mapping Operators

Čadík et al. [82, 83] presented a TMO evaluation study with subjective experiments using real scenes as a reference, similar to Yoshida et al.'s work [430, 431]. Both rating and ranking methodologies were employed. Furthermore, the collected data was fitted into different metrics. The main focus of the study was on some image and HVS attributes, including brightness or perceived luminance, contrast, reproduction of colors, and detail reproduction in bright and dark areas of the image. Ten participants took part in the two experiments, and 14 TMOs were employed: linear mapping, LCIS [374], revised Tumblin-Rushmeier [372], photographic tone reproduction [323], uniform rational quantization [341], histogram adjustment [215], fast bilateral filtering [118], trilateral filter [92], adaptive gain control [303], Ashikhmin's operator [27], gradient domain compression [129], contrast-based factor [406], adaptive logarithmic mapping [116], and spatially nonuniform scaling [90].

The first finding was that there is no statistical difference between the data of the two experiments, between rating and ranking. Therefore, the authors suggested that, for a perceptual comparison of TMOs, ranking without a reference is enough. The second finding was the good performance of global methods over local ones. This fact is in line with other studies, such as Ledda et al. [216], Akyüz et al. [17], Drago et al. [115], and Yoshida et al. [430, 431]. In the last part of the study, the relationship between the overall image quality and the four attributes was analyzed and fitted into parametric models for generating image metrics.

The study measured performances of a large number of TMOs. Furthermore, four important attributes of an image were measured, and not only the overall quality. However, the number of participants was small and the choice of scenes was very limited and did not cover other common real-world lighting conditions.

10.3.3 Evaluation of Video Tone Mappers

Eilertsen el al. [123] presented an evaluation of dynamic tone mappers used for tone mapping HDR video content. Their work involved the selection of video tone mappers and included two expert-driven pilot studies for selection of parameters and methods and a final pairwise comparison study with a more general population. The authors originally selected a total of 11 TMOs with a mixture of local and global methods, mostly from the literature and one Camera TMO based on the S-shaped tone curve of a Canon 500D DSLR camera. All three studies were conducted on a 24" 1920×1200 LCD display with a peak luminance of 200 cd/m^2 in a dim room. Six video sequences were chosen in total and reused throughout the experiments.

The first pilot study conducted by four domain experts was used to identify a set of parameters to be set for each method for those that did not have author-suggested parameters. Three sequences were used, distinct from those in the other experiments. The averaged parameters from those chosen were used for the subsequent experiments. The second pilot study involved five experts judging the original 11 dynamic TMOs on the criteria of overall brightness, contrast, saturation, consistency, temporal flickering, ghosting and noise via a rating procedure. The experts suggested the most objectionable issues corresponded to temporal characteristics, particularly flickering and ghosting; hence those methods that exhibited such artifacts consistently across scenes were eliminated from the subsequent experiment.

The selected seven methods were compared via a pairwise comparison methodology with the use of the quicksort technique [250] to reduce the number of comparisons. Eighteen participants took part in the experiment in which they selected the method considered best among the two viewed. While the results differed across scenes, three of the methods were more frequently preferred. These were the display adaptive TMO [242], the Irawan et al.'s TMO [182], and the camera specific one introduced above.

10.3.4 Evaluation of Video Tone Mappers on Small Screen Devices

The prevalence of portable devices has led to the consumption of a large amount of multimedia content on such devices. The emergence of HDR entails that such devices should eventually be able to support the display of such content and primarily, at the initial stages, this would be via the use of tone mappers. One characteristic of portable devices is their use in widely varying lighting conditions from bright sunlight to locations without any light. Melo et al. [257, 258] presented an evaluation of six video tone mappers on portable devices under varying lighting conditions in a controlled experiment. The study also investigated differences between the video tone mappers run on small screen devices and traditional LDR displays. Seven dynamic TMOs were investigated across seven scenes, two displays and three different lighting conditions. The TMOs were a within-participants independent variable while the lighting conditions and the displays were between-participants variables. The scenes were divided into two groups which were independent between participants. 180 participants took part in the experiment and were divided into groups of 15 that corresponded to one set of lighting, one set of display and a group of scenes. The procedure involved participants ranking the TMOs for each scene compared to a reference being shown on a SIM2 HDR display. The TMOs were displayed on either an LDR display or an Apple iPad representing a small screen device, depending on the participant group. Furthermore, each participant was exposed to one of three environmental lighting conditions. The lowest environment level was a dark indoor environment at 15 cd/m_2, medium was based on artificial lighting at 55 cd/m^2, and the final was representative of overcast outdoor conditions at 1,450 cd/m^2 produced via the use of photographic lights to maintain the controlled environment. Participants were asked to rank the six TMOs in order of preference.

Results showed that the display adaptive tone mapper [242] was considered best in the overall rankings when data was collapsed across all lighting conditions, displays and scenes. Boitard et al.'s method [68] was considered second, and Pattanaik et al.'s method [302] was third, with significant differences across all these. In the lighting specific analysis (across both displays), the display adaptive method was considered best for the low and medium environment levels while Pattanaik et al.'s method did best for the dark environments. When analyzing results across the small screen devices and LDR a similar pattern emerges with display adaptive being best for LDRs and small screen devices. Furthermore, a similar pattern to the overall occurs when analyzing across the different lighting conditions with display adaptive best for low and medium lighting conditions and Pattanaik et al.'s method best for bright environments. Congruence among participants was fairly high with Kendall's coefficient of concordance for agreement among participant ranking being consistently high and always significant. Results across scenes were quite diverse with different methods performing better for different scenes however the best method for each scene was always one of the top three reported for the overall results.

10.4   Experimental Evaluation of Expansion Operators

Expansion operators also benefit from the use of subjective evaluations since a number of significantly varying methods have been proposed. Akyüz et al. [17] compared the use of straightforward expansion methods, Banterle et al. [47] compared more complex published methods via pairwise comparisons against a reference, see below, and Masia et al. [251] performed a similar experiment on complex methods via rating comparing methods in a 2×2 grid, however, without any reference. Abebe et al. [1] presented a rating experiment designed to identify which expansion methods and color appearance models reproduced the best color appearance versus a ground truth HDR.

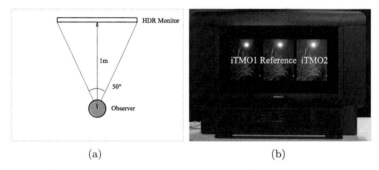

(a)   (b)

Figure 10.7. An example setup for the evaluation of expansion methods using an HDR monitor: (a) The diagram. (b) A photograph.

10.4.1   An Evaluation of Inverse Tone Mapping Techniques

Banterle et al. [47] proposed a subjective evaluation of expansion algorithms based on a pairwise comparisons methodology [103,216] using an HDR reference image displayed on the Brightside DR-37p HDR monitor [76]. Figure 10.7 illustrates the setup used for this evaluation. The study involved 24 participants, and five algorithms were tested: Banterle et al. [48,50] (**B**), Meylan et al. [262,263] (**M**), Wang et al. [402] (**W**), Rempel et al. [324] (**R**), and Akyüz et al. [17] (**A**). The study was divided into two experiments. The first one tested performances of various expansion algorithms for the recreation of eight HDR images starting from clipped ones. A participant had to choose the best picture in a pair that was closer to the reference: overall, in the dark areas, and in the bright ones. The second experiment investigated which expansion method performed best for recreating six HDR environment maps for IBL for three different materials: pure diffuse, pure specular, and glossy. Each participant had to choose the closest relit object (a teapot) to a reference relit object out of a pair of relit objects.

For the first experiment, the monotonically increasing functions **B**, **W**, and **R** that enhance contrast nonlinearly performed better overall and were grouped together in many of the results. The linear method **A**, and to a lesser extent **M**, performed worst overall, reflecting that for still images complex methods recreate HDR perceptually better.

For the second experiment, the diffuse results showed few differences. This is mostly due to the fact that rendering with IBL consists of evaluating an integral and during this integration small details may be lost. This is less true for perfectly mirror-like or high glossy materials. However, in these cases, details of the environment map reflected in the objects may be too small to be seen, as was shown by the large groupings in the results. For more complex environment maps, the previously found ranking was reverted. Overall, the results clearly showed that the operators that perform best, as with the first experiment, were the nonlinear operators.

This study showed that more advanced algorithms that cater for quantization errors introduced during expansion of an LDR image, such as **B**, **R**, and **W**, can perform better than simple techniques that apply single or multiple linear scale expansions, such as **A** and **M**. The more computationally expensive methods **B**, **R**, and **W** are better at recreating HDR than simple methods. Even if a linear scale can elicit an HDR experience in an observer, as shown by Akyüz et al. [17], it does not correctly reproduce the perception of the original HDR image.

10.5 HDR Compression Evaluation

Recently, with the widespread adoption of HDR, a significant number of HDR compression methods have been developed, see Chapter 8 and Chapter 9. This has led to several comparison studies of the proposed methods. Koz and Dufaux [203] provided a comparison between the two broad categories of single stream and dual stream methods discussed in Chapter 9. This study, however, is solely limited to objective evaluations with no participant involvement. Azimi et al. [34] evaluated via pairwise comparison the PQ method [270] with a generic tone-mapper followed by inverse tone mapper method. Hanhart et al. [162] presented a side-by-side evaluation of methods submitted to MPEG's call for evidence for HDR encoding in HEVC [236]. Mukharjee et al. [276] provided both subjective and objective evaluations comparing a number of methods discussed in Chapter 9 and is presented in further detail next.

10.5.1 Subjective and Objective Evaluation of HDR Video Compression

Mukharjee et al. [276] presented an objective and subjective evaluation of video compression methods for HDR content. The methods chosen for evaluation were, HDRv [245] (see Section 9.1), a video version of Ward and Simmons JPEG-HDR [410], Mantiuk et al.'s HDR-MPEG [244] (see Section 9.2), Lee and Kim's rate distortion method [220] (see Section 9.3), the Banterle et al.'s method [37] (see Section 9.8.1) and Garbas and Thoma's method [139] (see Section 9.4). The first step of the evaluation involved the comparison of the six methods across 39 video sequences, with an average dynamic range of 14 to 23 f-stops at 24 fps at full HD resolution (1920 × 1080). The sequences were graded for viewing on an HDR display with dynamic range between 10^{-3} to 4,000 cd/m^2. The method used for evaluation involved applying the HDR encoding method to the original HDR frames which generated a block of data in YUV format which was used as input to an H.264 encoder. Single stream methods were encoded at relevant bit depths as described in the original papers and the two stream methods were encoded as two 8-bit streams. 11 QP values per sequence were used to encode at different qualities. All were encoded at High 4:4:4 sub-sampling. The video streams were subsequently decoded from LDR to YUV blocks which were then converted back to HDR frames using the HDR decoding function of each of the six methods. The final HDR frames were subsequently compared with the originals using a number of objective metrics. Seven metrics were chosen for the objective evaluation across a range of traditional and HDR-specific metrics: PSNR, logPSNR, puPSNR, puSSIM, Weber-MSE, HDR-VDP, and HDR-VQM. Results demonstrate that in the overall the non-backwards compatible methods outperformed the other methods. The results of the perceptual metrics HDR-VDP, HDR-VQM, and puPSNR are very similar.

The subjective evaluation was conducted on a representative six of the original set of HDR sequences. The preparation of the content followed the same process as with the objective evaluation, except only two quality settings were chosen: 0.15 bpp (approximating streaming quality) and 0.75 bpp (approximating blu ray video quality at HD resolution). The procedure followed was a ranking one with comparison to a ground truth and with a hidden reference. Stimuli were viewed on a SIM2 HDR display with peak luminance of 4,000 cd/m^2, while the GUI for selection and rating was displayed on an LED display with peak rating of 300 cd/m^2. Participants were asked to view a ground truth sequence for each scene initially. Subsequently, they were asked to choose from seven possible sequences corresponding to the six methods and the hidden reference. They would then rank each one of those sequences in terms of similarity to the ground truth.

Sequences could be viewed as many times as desired. 64 participants took part in the study, divided equally into two groups. One group conducted all scenarios with the lower quality settings and the second group all scenarios with the higher quality settings. Results showed similarity with the objective scores with the single stream methods outperforming the other methods for both the high quality and low quality results. Kendall's coefficient of concordance, which reports agreements between participants, was very high at 0.783 for the low quality group and relatively high for the high quality group at 0.511. Both these results were statistically significant and entail a high level of agreement among the participants. The hidden reference outperformed the other methods but it was not considered significantly different from HDRV. The rest of the ordering consisted of Garbas and Thoma's method, the Banterle et al.'s method, the video version of JPEG-HDR, HDR-MPEG and the rate distortion method.

A comparison of results across the subjective and objective methods show significant correlations (using Spearman's Rho) between puPSNR, puSSIM, HDR-VDP and HDR-VQM and the subjective rankings with all correlations above 0.8. logPSNR and WeberMSE do reasonably well with correlations in the 0.6 to 0.76 range but are not considered statistically significant. Traditional PSNR is the worst performer with correlations of 0.371 for both high quality and low quality results.

10.6 Summary

Thorough and careful evaluation is a key part of any attempt to authentically simulate reality. Many methods and applications in HDR require the comparison of different methods. These include HDR compression, tone mapping and expansion operators. TMOs, in their various guises (static, dynamic) and platforms (HDR, LDR, mobile devices), require the same perceptual response as the real world scene. Not surprisingly, some TMOs were shown to be much better at simulating the real world than others. Similarly, as the results in this chapter show, certain expansion methods are able to create HDR content from LDR images more accurately than others. More recently, HDR compression methods have been evaluated to identify the methods most able to compress the large data requirements of HDR video without any significant loss in perceived quality.

Concurrently, the search for automated methods of evaluating different methods, without the complexities, both in terms of methodology and practicality, required by subjective experiments, has led to the development of a number of metrics for evaluating images and videos. Metrics offer a straightforward and objective means of comparing images via statistical methods and/or models of the HVS. Subjective experiments, on the other hand, use the HVS directly, to compare images. Although not limited by any restrictions of a computer model, these experiments also have certain issues. First, to provide meaningful results they may need to be run with a large number of participants. There is no such thing as a "normal" HVS and thus only by using large samples can any anomalies in participants' HVSs be sufficiently minimized. In addition, organizing the experiments is time-consuming and a large number of other factors such as participant fatigue/boredom, the environment in which the experiment is conducted, the scenes to be evaluated etc. have to be controlled to avoid bias. Metrics enable automation and the use of metrics can also be applied directly to methods, for example, to identify if further refinement for compression produces any noticeable differences. Although there has been substantial progress in modeling the HVS for the use in computational metrics, the complexity of the HVS has yet to be fully understood. As our understanding of the HVS increases, so too will the computational fidelity of computer metrics.

The bilateral filter [349, 368] is a non-linear filter that maintains strong edges while smoothing internal values. It is related to thetropic diffusion [52, 297, 307]. The filter for an image I is defined as

$$BF[I](\mathbf{x}, f_s, g_r) =$$

$$= \frac{1}{K[I](\mathbf{x}, f_s, g_r)} \sum_{\mathbf{y} \in \Omega(\mathbf{x})} I(\mathbf{y}) f_s(\|\mathbf{x} - \mathbf{y}\|) g_r(\|I(\mathbf{y}) - I(\mathbf{x})\|),$$

$$K[I](\mathbf{x}, f_s, g_r) = \sum_{\mathbf{y} \in \Omega(\mathbf{x})} f_s(\|\mathbf{x} - \mathbf{y}\|) g_r(\|I(\mathbf{y}) - I(\mathbf{x})\|), \qquad (A.1)$$

where f_s is the smoothing function for the distance term, g_r is the smoothing function for the intensity term, $\Omega(\mathbf{x})$ is the pixel neighborhood around pixel $\mathbf{x}$ that depends on f_s, and $K[I](\mathbf{x}, f_s, g_r)$ is the normalization term. In the standard notation, f_s's parameters have the $_s$ subscript, and g_r's parameters have the $_r$ subscript. Typically, f_s and g_s are Gaussian functions; i.e., $f(x) = \exp(-x^2/(2\sigma^2))$. Figure A.1 provides an example showing how the filter works on a step edge. An example of bilateral filtering on a natural image is shown in Figure A.2.

Furthermore, the bilateral filtering can be used for transferring edges from a source image to a target image by modifying Equation (A.1) as

$$CBF[I, J](\mathbf{x}, f_s, g_r) =$$

$$= \frac{1}{K[I, J](\mathbf{x}, f_s, g_r)} \sum_{\mathbf{y} \in \Omega(\mathbf{x})} I(\mathbf{y}) f_s(\|\mathbf{x} - \mathbf{y}\|) g_r(\|J(\mathbf{y}) - J(\mathbf{x})\|),$$

$$K[I, J](\mathbf{x}, f_s, g_r) = \sum_{\mathbf{y} \in \Omega(\mathbf{x})} f_s(\|\mathbf{x} - \mathbf{y}\|) g_r(\|J(\mathbf{y}) - J(\mathbf{x})\|),$$

where J is the image containing edges to be transferred to I. This version is called joint/cross bilateral filtering [199, 308]. When this version is used

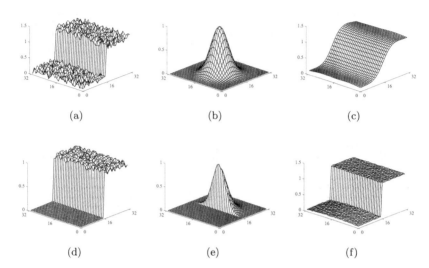

Figure A.1. An example of bilateral filtering: (a) An edge corrupted by additive Gaussian noise with $\sigma = 0.25$. (b) A spatial Gaussian kernel with $\sigma_s = 4$ evaluated at the central pixel. (c) Image in (a) filtered with the Gaussian kernel in (b); note that noise is removed but edges are smoothed causing a blurring effect. (d) An intensity Gaussian kernel with $\sigma_r = 0.25$ evaluated at the central pixel. (e) A spatial-intensity Gaussian kernel obtained by multiplying the kernel in (b) and (d), called the bilateral kernel. (f) Image in (a) filtered using the bilateral kernel; note that while edges are preserved, noise is removed.

Figure A.2. An example of bilateral filtering on a natural image: (a) The original image. (b) The image in (a) after applying a bilateral filter ($\sigma_s = 16$, and $\sigma_r = 0.1$). Note that strong edges are kept and small details are smoothed.

for up-sampling, it is called joint bilateral up-sampling [309]. This is a general technique to speed up computations of various tasks such as stereo matching, tone mapping, global illumination, etc. A task, which needs to

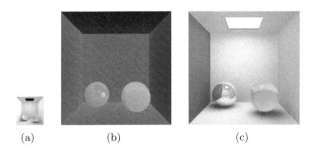

(a)   (b)   (c)

Figure A.3. An example of joint bilateral up-sampling for rendering: (a) The starting low resolution image representing indirect lighting. (b) A depth map used as an edge map. (c) The up-sampled image in (a) transferring edges of (b) with direct lighting added.

be sped up, is computed on a small scale (Figure A.3(a)) and then it is up-sampled (Figure A.3(c)) using the input resolution image or other feature maps (Figure A.3(b)).

The computational complexity of the bilateral filter is $\mathcal{O}(nk^2)$, where n is the number of pixels in the image and k is the radius of the spatial kernel. This computational complexity is relatively high for a filtering task, e.g., a medium-large filter (e.g.,120×120) can take up to 132 seconds for filtering an one megapixel HDR image (single precision floating point) on an Intel i7-860 running at 2.8 Ghz using a multi-thread implementation [41] using all four cores. Unfortunately, a bilateral filter is not separable (filtering first rows and then columns or viceversa) as a Gaussian filter because it is non-linear. Separation can work only for small kernels—for example, filters with a window less than 10 pixels [309]. For larger kernels, artifacts can be visible in the final image around edges. When kernels are not Gaussian, the bilateral filter can be accelerated by reorganizing computations because they are similar for neighbors of the current pixel [414]. A more general class of speed-up algorithms approximate the filtering using the so-called splat-blur-slice paradigm [2, 3, 89]. In such methods, the image is seen as a set of multidimensional points, where intensity or colors are coordinates similar to the spatial coordinates. Depending on the size of filtering kernels, points are projected onto a low resolution spatial data structure (splat) representing the multidimensional space of the image. At this point, filtering is computed on this data structure (blur), and then computed values are redistributed to the original pixels (slice). Another solution is to simply subsample $\Omega(\mathbf{x})$ using Poisson-disk distributed points [44]. This solution is straightforward to implement and fast, but it can create noise when the bilateral filter degenerates into a low pass filter (i.e., a g_r has a large support); although such case is rare.

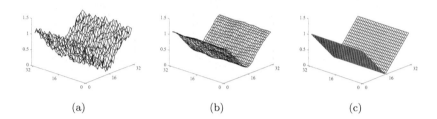

Figure A.4. A comparison between the bilateral and the trilateral filter: (a) The input noisy signal. (b) The signal in (a) smoothed using a bilateral filter. (c) The signal in (a) smoothed using a trilateral filter. Note that ramps and ridges are kept instead of being smoothed as happened in (b).

There are three main disadvantages of the bilateral filter. The first is that it smooths across sharp changes in gradients, blunting or blurring ramp edges and valley or ridge-like features; see Figure A.4(b). The second is that high gradient or high curvature regions are poorly smoothed because most nearby values are outliers that miss the filter window. The third is that it may include disjoint domains or high gradient regions.

To solve all these problems, the bilateral filter was extended by Choudhury and Tumblin [92]. The new filter, called the trilateral filter, tilts and skews the filter window using bilateral filtered image gradients ∇I_θ; see Figure A.4. This is achieved by using as input for g_r the proximity of the center value $I(\mathbf{x})$ to a plane instead of the Euclidean distance of coordinates. This plane is defined as

$$P(\mathbf{x}, \mathbf{y}) = I(\mathbf{x}) + \nabla I_\theta \cdot \mathbf{y},$$

where $\mathbf{x}$ are the pixel coordinates of the current pixel to filter, and $\mathbf{y}$ are the pixel coordinates of a sample in the filtering window. The only disadvantage of this filter is the high computational cost because two bilateral filters are needed to be calculated; one for gradients and another for the filtering of image values.

Recently, researchers have proposed several edge-aware filters to overcome bilateral filters' limitations. These filters are based on different ideas such as weighted least squares [126], wavelet [26,128], local linear model fitting [169], L0 gradient minimization [424], local statistics [189,357], multiscale pyramids [298], adaptive manifolds [140], L1 image transform [62], etc. All these methods provide edge preserving results, and they can be applied iteratively in a multi-scale fashion.

B
Practical Color Spaces

This appendix briefly describes some color space conversion formulas that are typically used for converting between color spaces. These can be used in both an LDR and HDR imaging context. All these color space conversions are implemented in the HDR Toolbox and their MATLAB implementations may be found in the folder `ColorSpaces`.

CIE $L^\star a^\star b^\star$ (or CIE LAB) is a device-independent color space, which was introduced by the CIE to meet the requirement of having a color space that is perceptually uniform. CIE $L^\star a^\star b^\star$ is based upon CIE XYZ and is another attempt to linearize the perceptibility of unit vector color differences. The $L^\star$ component represents the lightness of a color and has values in the range $[0, 100]$. $a^\star$ represents its position between green ($a^\star < 0$) and magenta ($a^\star \geq 0$), $b^\star$ represents its position between blue ($b^\star < 0$) and yellow ($b^\star \geq 0$). Although, CIE $L^\star a^\star b^\star$ is a device-independent color space, it suffers from being quite unintuitive. Although the $L^\star$ parameter has a good correlation with lightness, $a^\star$ and $b^\star$ do not have a straightforward mapping for editing.

The conversion from a CIE XYZ value to a CIE $L^\star a^\star b^\star$ value is defined as

$$\begin{bmatrix} L^\star \\ a^\star \\ b^\star \end{bmatrix} = \begin{bmatrix} 116 \cdot f(Y/Y_n) \\ 500 \cdot \big(f(X/X_n) - f(Y/Y_n)\big) \\ 200 \cdot \big(f(Y/Y_n) - f(Z/Z_n)\big) \end{bmatrix},$$

where $[X_n, Y_n, Z_n]$ is the color value of the reference white point, which is $[0.95047, 1.0, 1.08883]$ for a $D65$ illuminant with Y normalization. f is defined as

$$f(x) = \begin{cases} x^{\frac{1}{3}} & \text{if } x > t^3, \\ \frac{x}{3t^2} + \frac{4}{29} & \text{otherwise.} \end{cases} \quad \text{where } t = \frac{6}{29}$$

The MATLAB function that implements this color space conversion and its inverse is called `ConvertXYZtoCIELab.m`.

CIE LCh is simply the cylindrical version of the CIE $L^\star a^\star b^\star$ color space. In this case, L is the lightness (similar to $L^\star$), C is the chroma or distance from the center of the cylinder (i.e., saturation), and h is the hue angle in the color wheel.

The conversion from a CIE $L^\star a^\star b^\star$ value to a CIE LCh value is defined as

$$\begin{bmatrix} L \\ C \\ h \end{bmatrix} = \begin{bmatrix} L^\star \\ \sqrt{(a^\star)^2 + (b^\star)^2} \\ \tan^{-1}(a^\star / b^\star) \end{bmatrix}.$$

The MATLAB function that implements this color space conversion and its inverse is called `ConvertXYZtoCIELCh.m`.

CIE $L^\star u^\star v^\star$ (or CIE Luv) is a device-independent color space, which was introduced by the CIE to meet the requirement of having a color space that is perceptually uniform. As CIE $L^\star a^\star b^\star$, CIE $L^\star u^\star v^\star$ is based upon CIE XYZ. The $L^\star$ component represents the lightness of a color and has values in the range $[0, 100]$. $u^\star$ represents its position between blue ($u^\star \geq 0$) and red ($u^\star > 0$), $v^\star$ represents its position between magenta ($v^\star < 0$) and yellow ($v^\star \geq 0$).

The conversion from a CIE XYZ value to a CIE $L^\star u^\star v^\star$ value is defined as

$$\begin{bmatrix} L^\star \\ u^\star \\ v^\star \end{bmatrix} = \begin{bmatrix} f(Y/Y_n) \\ 13 \cdot L^\star \cdot (u' - u'_n) \\ 13 \cdot L^\star \cdot (v' - v'_n) \end{bmatrix},$$

where $[X_n, Y_n, Z_n]$ and (u'_n, v'_n) are, respectively, the color value of the reference white point and its chromaticity coordinates. (u', v') (the chromaticity coordinates) are defined as

$$u' = \frac{4X}{X + 15Y + 3Z}, \quad v' = \frac{9Y}{X + 15Y + 3Z}.$$

f is defined as

$$f(x) = \begin{cases} 116 \cdot x^{\frac{1}{3}} - 16 & \text{if } x > \left(\frac{6}{29}\right)^3, \\ \left(\frac{29}{3}\right)^3 \cdot x^3 & \text{otherwise.} \end{cases} \tag{B.1}$$

The MATLAB function that implements this color space conversion and its inverse is called `ConvertXYZtoCIELUV.m`.

LMS is a color space that represents the response of the three types of cones of the human eye for chromatic vision, see Chapter 1.1.2. L is the response of the long wavelengths (red color), M is the response of the medium wavelengths (green color), and S is the response of the short wavelength (blue color).

The conversion from a CIE XYZ value to a LMS value is defined as

$$
\begin{bmatrix} L \\ M \\ S \end{bmatrix} = \mathbf{M}_{LMS} \cdot \begin{bmatrix} X \\ Y \\ Z \end{bmatrix} \text{ where } \mathbf{M}_{LMS} = \begin{bmatrix} 0.4002 & 0.7075 & -0.0807 \\ -0.2280 & 1.1500 & 0.0612 \\ 0.0000 & 0.0000 & 0.9184 \end{bmatrix}.
$$

Note that $\mathbf{M}_{LMS}$ is the Hunt-Pointer-Estevez matrix for a D65 illuminant. The MATLAB function that implements this color space conversion and its inverse is called `ConvertXYZtoLMS.m`.

IPT is a color space that builds upon LMS. It provides hue uniformity, making it a desirable color space for gamut mapping [120].

The conversion from an LMS value to an IPT value is defined as

$$
\begin{bmatrix} I \\ P \\ T \end{bmatrix} = \mathbf{M}_{IPT} \cdot \begin{bmatrix} L' \\ M' \\ S' \end{bmatrix} \text{ where } \mathbf{M}_{IPT} = \begin{bmatrix} 0.4000 & 0.4000 & 0.2000 \\ 4.4550 & -4.8510 & 0.3960 \\ 0.8056 & 0.3572 & -1.1628 \end{bmatrix},
$$

where $[L', M', S']$ are the cones' responses after signal compression. This is defined by a non-linear function as

$$
\begin{bmatrix} L' \\ M' \\ S' \end{bmatrix} = \begin{bmatrix} f(L) \\ f(M) \\ f(S) \end{bmatrix} \text{ where } f(x) = \begin{cases} x^{0.43} & \text{if } x \geq 0, \\ -(-x)^{0.43} & \text{otherwise.} \end{cases}
$$

The MATLAB function that implements this color space conversion and its inverse is called `ConvertXYZtoIPT.m`.

A Brief Overview of the MATLAB HDR Toolbox

This appendix describes how to use the HDR Toolbox presented in this book. The HDR Toolbox is available on the book website[1] or on GitHub[2]. The HDR Toolbox is self-contained, although it uses some functions from the Image Processing Toolbox by Mathworks [255]. Note that HDR built-in functions of MATLAB such as `hdrread.m`, `hdrwrite.m`, and `tonemap.m`, need MATLAB version 2008b or above. Furthermore, MATLAB version 2010a and the respective Image Processing Toolbox are needed for compressing images for HDR JPEG2000. The initial process for handling HDR images/frames in MATLAB is to either load them or create them.

The HDR Toolbox provides the function `hdrimread.m` to read HDR images. This function takes as input a MATLAB string and outputs an `m-n-3` matrix. This function can be run as:

```
>> img = hdrimread('Bottles\_Small.hdr');
```

Note that `hdrimread.m` can read *portable float maps* files (.pfm), RLE compressed and uncompressed *radiance* files (.hdr/.pic), and *OpenEXR* files (.exr) using the TinyEXR library[3]. Furthermore, this function can read all LDR file formats that are supported natively by MATLAB and it will automatically normalize them to the range $[0, 1]$ with double precision. MATLAB from version 2008b onward provides support for reading *radiance* files both compressed (using run-length encoding) and uncompressed. An example of how to use this function for loading 'Bottles_Small.hdr' is:

```
>> img = hdrread('Bottles\_Small.hdr');
```

---

[1] http://www.advancedhdrbook.com/
[2] https://github.com/banterle/HDR_Toolbox
[3] https://github.com/syoyo/tinyexr

In order to generate HDR images, the HDR Toolbox provides the function BuildHDR.m to produce an HDR image based on a sequence of LDR images. This function takes as inputs a stack of LDR images (stack), which is a 4D matrix, an array with the exposure values of each LDR image (stack_exposure), the type of CRF (lin_type) and its data (lin_fun), the weighting function (weight_type), and the merging domain (merge_type). The output of the function is an HDR image, imgOut stored as a m-n-3 matrix in the case of an *RGB* image. To read an HDR stack and its exposure values, the functions ReadLDRStack and ReadLDRStackInfo can be respectively employed. To create a CRF from a stack and its exposure value, the function DebevecCRF [111] may be used; see the folder Generation for other functions such as MitsunagaNayarCRF.m [271], RobertsonCRF.m [332], etc. An example of a full set of operations to be launched for creating an HDR image is:

```
>> [stack, norm_value] = ReadLDRStack('stack', 'jpg', 1);
>> stack_exposure = ReadLDRStackInfo('stack', 'jpg');
>> [lin_fun, ~] = DebevecCRF(stack, stack_exposure);
>> imgOut = BuildHDR(stack, stack_exposure, 'LUT', ...
>> lin_fun, 'Deb97', 'log');
```

Note that the LDR stack is merged in the logarithmic domain ('log') to reduce camera noise, and Debevec and Malik's weighting function [111] is used ('Deb97').

Figure C.1. The Bottles HDR image gamma corrected, with a gamma value of 2.2 for display at f-stop 0.

Once images are either loaded into memory or generated, a useful operation is to visualize them. A straightforward operation that allows single-exposure images to be shown is `GammaTMO.m`. This function applies gamma correction to an HDR image at a given f-stop value and displays it on the screen. Note that values are clamped between [0, 1]. For example, if an HDR image needs to be displayed at f-stop 0, with a gamma value of 2.2, the following command needs to be typed on the MATLAB console:

```
>> GammaTMO(img, 2.2, 0, 1);
```

The result of this operation can be seen in Figure C.1.

To store a gamma-corrected exposure into a matrix without viewing it, the visualization flag (`TMO_view`) needs to be set to 0:

```
>> imgOut = GammaTMO(img, 2.2, 0, 0);
```

Gamma-corrected single-exposure images are a straightforward way to view HDR images, but they do not permit the large range of luminance in an HDR image to be properly viewed. The HDR Toolbox provides several TMOs that can be used to compress the luminance in order to be visualized on an LDR monitor. For example, to tone map an image using Drago et al.'s operator [116], the `DragoTMO.m` function is used and the image is saved into a temporary image.

Figure C.2. The Bottles HDR image tone mapped with Drago's TMO [116].

Then, this image is visualized using the `GammaTMO.m` function as shown before:

```
>> imgTMO = DragoTMO(img);
>> GammaTMO(imgTMO, 2.2, 0, 1);
```

The result of this can be seen in Figure C.2.

If the tone mapping is not satisfactory, its parameters can be changed. Each TMO can be queried with the command `help` that can be used to understand which parameters can be set and what these parameters do. Help can be called using the command `help nameFunction`:

```
>> help DragoTMO;
>> imgTMO = DragoTMO(img, 100.0, 0.5);
>> GammaTMO(imgTMO, 2.2, 0, 1);
```

Figure C.3 shows the resulting image.

Figure C.3. The Bottles HDR image tone mapped with Drago's TMO [116] changing the bias parameter to 0.5.

Furthermore, if colors in tone mapped images are too saturated due to the fact that the range is compressed, they can be adjusted using the `ColorCorrection.m` function. This function increases or decreases saturation in the image. In the case of Figure C.3, the image needs to be desaturated. To achieve this goal, the color correction parameter, `cc_s` needs to be set in the range $(0, 1)$:

```
>> imgTMO = DragoTMO(img, 100.0, 0.5);
>> imgTMOCor = ColorCorrection(imgTMO,1.0, 0.4);
>> GammaTMO(imgTMOCor, 2.2, 0, 1);
```

Figure C.4 shows the results of these sequence of commands.

Expansion operators provide a tool for creating HDR content from LDR images. Similarly to TMOs, expansion operators' parameters can be queried via the help function.

Figure C.4. The Bottles HDR image tone mapped with Drago's TMO [116] followed by a color correction step.

For example, the expansion of an LDR image (e.g., 'Venice01.png' provided in the HDR Toolbox) using the Kovaleski and Oliveira's operator [202] is straightforward. The first step is to load the LDR image using the function (ldrimread.m, which casts 8-bit images into double precision and maps them in the range [0, 1]. Then, the operator is queried with help to understand the parameters required. Finally, the operator is executed by calling the function KovaleskiOliveiraEO.m:

```
>> img = ldrimread('Venice01.png');
>> help KovaleskiOliveiraEO
>> imgEXP = KovaleskiOliveiraEO(img, 'image', 150, ...
>>   0.1, 3000, 0.15, 2.2);
```

EOs can reduce saturation during expansion, the opposite case of what happens during tone mapping. In this case, the ColorCorrection.m can be used to increase the saturation using a greater than one correction value such as

```
>> imgEXPCOR = ColorCorrection(imgEXP, 1.4);
```

Loaded, tone mapped, and expanded images may need to be stored in secondary storage such as a hard drive. The HDR Toolbox has a native function to write .pfm, .hdr, and .exr files hdrimwrite.m. For instance, to write an image as an .exr file, the function hdrimwrite.m is called as

```
>> hdrimwrite(imgEXP, 'out.exr');
```

Note that MATLAB also provides a native function, hdrwrite.m, to write .hdr files:

```
>> hdrwrite(imgEXP, 'out.hdr');
```

The HDR Toolbox provides other functions for manipulating HDR images, including bilateral decomposition, histogram calculation, merging of LDR images into HDR ones, light source sampling, etc. All these functions are straightforward to use. Each function's help is a useful tool for providing more details on the function's description, parameters, and outputs.

Bibliography

[1] Mekides Assefa Abebe, Tania Pouli, and Jonathan Kervec. Evaluating the Color Fidelity of ITMOs and HDR Color Appearance Models. *ACM Trans. Appl. Percept.*, 12(4):14:1–14:16, September 2015.

[2] Andrew Adams, Jongmin Baek, and Myers Abraham Davis. Fast high-dimensional filtering using the permutohedral lattice. *Computer Graphics Forum*, 29(2):753–762, 2010.

[3] Andrew Adams, Natasha Gelfand, Jennifer Dolson, and Marc Levoy. Gaussian kd-trees for fast high-dimensional filtering. *ACM Trans. Graph.*, 28(3):1–12, 2009.

[4] Ansel Adams. *The Print: The Ansel Adams Photography Series 3*. Little, Brown and Company, Cambridge, MA, USA, 1981.

[5] Edward H. Adelson. Saturation and adaptation in the rod system. *Vision Research*, 22:1299–1312, 1982.

[6] Edward H. Adelson and James R. Bergen. The Plenoptic Function and the Elements of Early Vision. In *Computational Models of Visual Processing*, pages 3–20, Cambridge, MA, USA, 1991. MIT Press.

[7] Adobe. Adobe PhotoShop. Available at http://www.adobe.com/products/photoshop.html, November 2016.

[8] Sameer Agarwal, Ravi Ramamoorthi, Serge Belongie, and Henrik Wann Jensen. Structured importance sampling of environment maps. *ACM Trans. on Graph.*, 22(3):605–612, 2003.

[9] Aseem Agarwala, Mira Dontcheva, Maneesh Agrawala, Steven Drucker, Alex Colburn, Brian Curless, David Salesin, and Michael Cohen. Interactive digital photomontage. *ACM Trans. Graph.*, 23(3):294–302, 2004.

[10] Manoj Aggarwal and Narendra Ahuja. Split aperture imaging for high dynamic range. *Int. J. Comput. Vision*, 58(1):7–17, 2004.

[11] Cecilia Aguerrebere, Andrés Almansa, Yann Gousseau, Julie Delon, and Pablo Musé. Single shot high dynamic range imaging using piecewise linear estimators. In *Proceedings International Conference on Computational Photography*. IEEE Computer Society, 2014.

[12] Damla Ezgi Akçora, Francesco Banterle, Massimiliano Corsini, Ahmet Oguz Akyuz, and Roberto Scopigno. Practical-hdr: A simple and effective method for merging high dynamic range videos. In Kenny Mitchell, editor, *The 13th European Conference on Visual Media Production (CVMP)*. ACM, ACM Press, December 2016.

[13] Tomas Akenine-Möller, Eric Haines, and Natty Hoffman. *Real-Time Rendering, Third Edition*. A K Peters, Ltd., Natick, MA, USA, 2008.

[14] Ahmet Oğuz Akyüz, M Levent Eksert, and M Selin Aydın. An evaluation of image reproduction algorithms for high contrast scenes on large and small screen display devices. *Computers & Graphics*, 37(7):885–895, 2013.

[15] Ahmet Oğuz Akyüz. High dynamic range imaging pipeline on the gpu. *J. Real-Time Image Process.*, 10(2):273–287, June 2015.

[16] Ahmet Oğuz Akyüz and Erik Reinhard. Noise reduction in high dynamic range imaging. *Journal of Visual Communication and Image Representation*, 18(5):366–376, 2007.

[17] Ahmet Oğuz Akyüz, Roland Fleming, Bernhard E. Riecke, Erik Reinhard, and Heinrich H. Bülthoff. Do hdr displays support ldr content?: A psychophysical evaluation. *ACM Trans. Graph.*, 26(3):38, 2007.

[18] Ahmet Oğuz Akyüz and Erik Reinhard. Color appearance in high-dynamic-range imaging. *Journal of Electronic Imaging*, 15(3):033001–1–033001–12, July-September 2006.

[19] Xiaobo An and Fabio Pellacini. AppProp: all-pairs appearance-space edit propagation. *ACM Trans. Graph.*, 27(3):40:1–40:9, August 2008.

[20] ArriAlexa. Alexa cameras. Available at http://www.arri.com/camera/alexa/, October 2016.

[21] Alessandro Artusi, Ahmet Oğuz Akyüz, Benjamin Roch, Despina Michael, Yiorgos Chrysanthou, and Alan Chalmers. Selective local tone mapping. In *International Conference on Image Processing (ICIP), September 15–18*, pages 2309–2313. IEEE, 2013.

[22] Alessandro Artusi, Jiří Bittner, Michael Wimmer, and Alexander Wilkie. Delivering interactivity to complex tone mapping operators. In *Proceedings of the 14th Eurographics Workshop on Rendering*, EGRW '03, pages 38–44, Aire-la-Ville, Switzerland, 2003. Eurographics Association.

[23] Alessandro Artusi, Rafał Mantiuk, Thomas Richter, Philippe Hanhart, Pavel Korshunov, Massimiliano Agostinelli, Arkady Ten, and Touradj Ebrahimi. Overview and evaluation of the JPEG-XT HDR image compression standard. *Journal of Real-Time Image Processing*, pages 1–16, 2015.

[24] Alessandro Artusi, Rafał Mantiuk, Thomas Richter, Pavel Korshunov, Philippe Hanhart, Touradj Ebrahimi, and Massimiliano Agostinelli. JPEG XT: A Compression Standard for HDR and WCG Images [Standards in a Nutshell]. *IEEE Signal Processing Magazine*, 33:118–124, 2016.

[25] Alessandro Artusi, Despina Michael, Benjamin Roch, Yiorgos Chrysanthou, and Alan Chalmers. A selective approach for tone mapping high dynamic range content. In *SIGGRAPH Asia 2011 Posters*, SA '11, pages 50:1–50:1, New York, NY, USA, 2011. ACM.

[26] Alessandro Artusi, Zhuo Su, Zongwei Zhang, Dimitris Drikakis, and Xiaonan Luo. High-order wavelet reconstruction for multi-scale edge aware tone mapping. *Comput. Graph.*, 45(C):51–63, December 2014.

[27] Michael Ashikhmin. A tone mapping algorithm for high contrast images. In *EGRW '02: Proceedings of the 13th Eurographics Workshop on Rendering*, pages 145–156, Aire-la-Ville, Switzerland, 2002. Eurographics Association.

[28] Michael Ashikhmin and Jay Goyal. A reality check for tone-mapping operators. *ACM Transaction on Applied Perception*, 3(4):399–411, 2006.

[29] Mekides Assefa, Tania Pouli, Jonathan Kervec, and Mohamed-Chaker Larabi. Correction of over-exposure using color channel correlations. In *IEEE Global Conference on Signal and Information Processing (Global-SIP), December 3–5*, pages 1078–1082, 2014.

[30] Tunç Aydın, Rafał Mantiuk, and Hans-Peter Seidel. Extending quality metrics to full luminance range images. In *Electronic Imaging 2008*, pages 68060B–68060B. International Society for Optics and Photonics, 2008.

[31] Tunç Ozan Aydın, Rafał Mantiuk, Karol Myszkowski, and Hans-Peter Seidel. Dynamic range independent image quality assessment. *ACM Trans. Graph.*, 27(3):1–10, 2008.

[32] Tunç Ozan Aydın, Nikolce Stefanoski, Simone Croci, Markus Gross, and Aljoscha Smolic. Temporally coherent local tone mapping of hdr video. *ACM Trans. Graph.*, 33(6):196:1–196:13, November 2014.

[33] Tunç Ozan Aydın, Martin Čadík, Karol Myszkowski, and Hans-Peter Seidel. Video quality assessment for computer graphics applications. *ACM Trans. Graph.*, 29(6):161:1–161:12, December 2010.

[34] Maryam Azimi, Ronan Boitard, Basak Oztas, Stelios Ploumis, Hamid Reza Tohidypour, Mahsa T Pourazad, and Panos Nasiopoulos. Compression efficiency of hdr/ldr content. In *Quality of Multimedia Experience (QoMEX), 2015 Seventh International Workshop on*, pages 1–6. IEEE Computer Society, 2015.

[35] Abhishek Badki, Nima Khademi Kalantari, and Pradeep Sen. Robust Radiometric Calibration for Dynamic Scenes in the Wild. In *Proceedings of the International Conference on Computational Photography (ICCP)*. IEEE Computer Society, April 2015.

[36] Francesco Banterle. *Inverse Tone Mapping*. PhD thesis, University of Warwick, 2009.

[37] Francesco Banterle, Alessandro Artusi, Kurt Debattista, Patrick Ledda, Alan Chalmers, Gavin J. Edwards, and Gerhard Bonnet. Hdr video data compression devices and methods. Available at https://register.epo.org/application?number=EP08012452&tab=main, July 2008. EP Patent App. EP20,080,012,452.

[38] Francesco Banterle, Alessandro Artusi, Elena Sikudova, Thomas Edward William Bashford-Rogers, Patrick Ledda, Marina Bloj, and Alan Chalmers. Dynamic range compression by differential zone mapping based on psychophysical experiments. In *ACM Symposium on Applied Perception (SAP)*. ACM Press, August 2012.

[39] Francesco Banterle, Alessandro Artusi, Elena Sikudova, Patrick Ledda, Thomas Edward William Bashford-Rogers, Alan Chalmers, and Marina Bloj. Mixing tone mapping operators on the gpu by differential zone mapping based on psychophysical experiments. *Elsevier Image Communication*, 48(October):50–62, October 2016. http://dx.doi.org/10.1016/j.image.2016.09.004.

[40] Francesco Banterle and Paolo Banterle. æ-hdr: An automatic exposure framework for high dynamic range content. In *ACM SIGGRAPH 2008 Posters*, SIGGRAPH '08, pages 58:1–58:1, New York, NY, USA, 2008. ACM.

[41] Francesco Banterle and Luca Benedetti. PICCANTE: An Open and Portable Library for HDR Imaging. http://vcg.isti.cnr.it/piccante, 2014.

[42] Francesco Banterle, Marco Callieri, Matteo Dellepiane, Massimiliano Corsini, Fabio Pellacini, and Roberto Scopigno. EnvyDepth: An Interface for Recovering Local Natural Illumination from Environment Maps. *Computer Graphics Forum (Proc. of Pacific Graphics 2013)*, 32(7):411–420, 2013.

[43] Francesco Banterle, Alan Chalmers, and Roberto Scopigno. Real-time high fidelity inverse tone mapping for low dynamic range content. In Olga Sorkine Miguel A. Otaduy, editor, *Eurographics 2013 Short Papers*, pages 41–44. Eurographics Association, May 2013.

[44] Francesco Banterle, Massimiliano Corsini, Paolo Cignoni, and Roberto Scopigno. A low-memory, straightforward and fast bilateral filter through subsampling in spatial domain. *Computer Graphics Forum*, 31(1):19–32, February 2012.

[45] Francesco Banterle, Kurt Debattista, Patrick Ledda, and Alan Chalmers. A gpu-friendly method for high dynamic range texture compression using inverse tone mapping. In *GI '08: Proceedings of Graphics Interface 2008*, pages 41–48, Toronto, Ontario, Canada, 2008. Canadian Information Processing Society.

[46] Francesco Banterle, Matteo Dellepiane, and Roberto Scopigno. Enhancement of low dynamic range videos using high dynamic range backgrounds. In RafałMantiuk and Erik Reinhard, editors, *Annex to the Conference Proceedings: Areas Papers on HDR*, Eurographics Area Papers on HDR, pages 57–62. Eurographics Association, April 2011.

[47] Francesco Banterle, Patrick Ledda, Kurt Debattista, Alessandro Artusi, Marina Bloj, and Alan Chalmers. A psychophysical evaluation of inverse tone mapping techniques. *Computer Graphics Forum*, 28(1):13–25, March 2009.

[48] Francesco Banterle, Patrick Ledda, Kurt Debattista, and Alan Chalmers. Inverse tone mapping. In *GRAPHITE '06: Proceedings of the 4th International Conference on Computer Graphics and Interactive Techniques in Australasia and Southeast Asia*, pages 349–356, New York, NY, USA, 2006. ACM.

[49] Francesco Banterle, Patrick Ledda, Kurt Debattista, and Alan Chalmers. Expanding low dynamic range videos for high dynamic range applications. In *SCCG '08: Proceedings of the 4th Spring Conference on Computer Graphics*, pages 349–356, New York, NY, USA, 2008. ACM.

[50] Francesco Banterle, Patrick Ledda, Kurt Debattista, Alan Chalmers, and Marina Bloj. A framework for inverse tone mapping. *The Visual Computer*, 23(7):467–478, 2007.

[51] Francesco Banterle and Roberto Scopigno. BoostHDR: A novel backward-compatible method for HDR images. In Andrew G. Tescher, editor, *Proceedings of SPIE Volume 8499: Applications of Digital Image Processing XXXV*, volume 8499. SPIE, August 2012.

[52] Danny Barash. A fundamental relationship between bilateral filtering, adaptive smoothing, and the nonlinear diffusion equation. *IEEE Trans. Pattern Anal. Mach. Intell.*, 24(6):844–847, June 2002.

[53] Connelly Barnes, Eli Shechtman, Adam Finkelstein, and Dan B Goldman. Patchmatch: A randomized correspondence algorithm for structural image editing. *ACM Trans. Graph.*, 28(3):24:1–24:11, July 2009.

[54] Rasmus Barringer and Tomas Akenine-Möller. Dynamic ray stream traversal. *ACM Trans. Graph.*, 33(4):151:1–151:9, July 2014.

[55] Peter GJ Barten. Formula for the contrast sensitivity of the human eye. In *Electronic Imaging 2004*, pages 231–238. International Society for Optics and Photonics, 2003.

[56] Thomas Bashford-Rogers, Kurt Debattista, and Alan Chalmers. Importance driven environment map sampling. *IEEE Trans. Vis. Comput. Graph.*, 20(6):907–918, 2014.

[57] Eric P. Bennett and Leonard McMillan. Video enhancement using per-pixel virtual exposures. *ACM Trans. Graph.*, 24(3):845–852, July 2005.

[58] Alexandre Benoit, David Alleysson, Jeanny Herault, and Patrick Le Callet. Spatio-temporal tone mapping operator based on a retina model. In *International Workshop on Computational Color Imaging (CCIW)*, pages 12–22, March 2009.

[59] Marcelo Bertalmio, Liminita A. Vese, Gullermo Sapiro, and Stanley J. Osher. Simultaneous structure and texture image in painting. *IEEE Transactions on Image Processing*, 12(8):882–889, August 2003.

[60] Pravin Bhat, Brian Curless, Michael Cohen, and Larry Zitnick. Fourier analysis of the 2d screened poisson equation for gradient domain problems. In *10th European Conference on Computer Vision (ECCV)*, pages 114–128, October 2008.

[61] Pravin Bhat, C. Lawrence Zitnick, Noah Snavely, Aseem Agarwala, Maneesh Agrawala, Brian Curless, Michael Cohen, and Sing Bing Kang. Using photographs to enhance videos of a static scene. In Jan Kautz and Sumanta Pattanaik, editors, *Rendering Techniques 2007 (Proceedings Eurographics Symposium on Rendering)*, pages 327–338. Eurographics, June 2007.

[62] Sai Bi, Xiaoguang Han, and Yizhou Yu. An l1 image transform for edge-preserving smoothing and scene-level intrinsic decomposition. *ACM Trans. Graph.*, 34(4):78:1–78:12, July 2015.

[63] Oliver Bimber and Daisuke Iwai. Superimposing dynamic range. *ACM Trans. Graph.*, 27(5):150:1–150:8, December 2008.

[64] Cambodge Bist, Rémi Cozot, Gérard Madec, and Xavier Ducloux. Style aware tone expansion for hdr displays. In *Proceedings of Graphics Interface 2016*, GI 2016, pages 57–63. Canadian Human-Computer Communications Society/Société canadienne du dialogue humain-machine, 2016.

[65] Benedikt Bitterli, Jan Novák, and Wojciech Jarosz. Portal-masked environment map sampling. *Computer Graphics Forum*, 34(4):13–19, 2015.

[66] James F. Blinn and Martin E. Newell. Texture and reflection in computer generated images. In *SIGGRAPH '76: Proceedings of the 3rd Annual Conference on Computer Graphics and Interactive Techniques*, pages 266–266, New York, NY, USA, 1976. ACM.

[67] R. Boitard, M. T. Pourazad, P. Nasiopoulos, and J. Slevinski. Demystifying high-dynamic-range technology: A new evolution in digital media. *IEEE Consumer Electronics Magazine*, 4(4):72–86, 2015.

[68] Ronan Boitard, Kadi Bouatouch, Remi Cozot, Dominique Thoreau, and Adrien Gruson. Temporal coherency for video tone mapping. In *Proc. SPIE 8499, Applications of Digital Image Processing XXXV*, volume 8499, pages 84990D–84990D–10. SPIE, 2012.

[69] Ronan Boitard, Remi Cozot, Dominique Thoreau, and Kadi Bouatouch. Survey of Temporal Brightness Artifacts in Video Tone Mapping. In *Second International Conference and SME Workshop on HDR Imaging*, March 2014.

[70] Ronan Boitard, Remi Cozot, Dominique Thoreau, and Kadi Bouatouch. Zonal brightness coherency for video tone mapping. *Image Communication*, 29(2):229–246, February 2014.

[71] Jennifer Bonnard, Céline Loscos, Gilles Valette, Jean-Michel Nourrit, and Laurent Lucas. High-dynamic range video acquisition with a multiview camera. In *Photonics Europe 2012 : Optics, Photonics and Digital Technologies for Multimedia Applications*, Brussel, Belgium, April 2012. SPIE.

[72] Nicolas Bonneel, James Tompkin, Kalyan Sunkavalli, Deqing Sun, Sylvain Paris, and Hanspeter Pfister. Blind video temporal consistency. *ACM Trans. Graph.*, 34(6):196:1–196:9, October 2015.

[73] Tim Borer and Andrew Cotton. A display-independent high dynamic range television system. *SMPTE Motion Imaging Journal*, 125(4):50–56, 2016.

[74] Alberto Boschetti, Nicola Adami, Riccardo Leonardi, and Masahiro Okuda. Flexible and effective high dynamic range image coding. In *Proceedings of 2010 IEEE 17th International Conference on Image Processing*, pages 3145–3148, 2010.

[75] Yuri Boykov, Olga Veksler, and Ramin Zabih. Fast approximate energy minimization via graph cuts. *IEEE Trans. Pattern Anal. Mach. Intell.*, 23(11):1222–1239, November 2001.

[76] Technologies BrightSide. DR37P HDR display. Available at https://en. wikipedia.org/wiki/BrightSide_Technologies, January 2006.

[77] Michael Brown, Aditi Majumder, and Ruigang Yang. Camera-based calibration techniques for seamless multiprojector displays. *IEEE Transactions on Visualization and Computer Graphics*, 11(2):193–206, March 2005.

[78] Neil D. B. Bruce. Special section on graphics interface: Expoblend: Information preserving exposure blending based on normalized log-domain entropy. *Comput. Graph.*, 39:12–23, April 2014.

[79] Antoni Buades, Bartomeu Coll, and Jean-Michel Morel. The staircasing effect in neighborhood filters and its solution. *IEEE Trans. Image Processing*, 15(6):1499–1505, 2006.

[80] David Burke, Abhijeet Ghosh, and Wolfgang Heidrich. Bidirectional importance sampling for direct illumination. In *Rendering Techniques 2005 Eurographics Symposium on Rendering*, pages 147–156, Aire-la-Ville, Switzerland, 2005. Eurographics Association.

[81] Peter J. Burt and Edward H. Adelson. The Laplacian pyramid as a compact image code. In *Readings in Computer Vision: Issues, Problems, Principles, and Paradigms*, pages 671–679. Morgan Kaufmann Publishers Inc., 1987.

[82] Martin Čadík, Michael Wimmer, Laszlo Neumann, and Alessandro Artusi. Image attributes and quality for evaluation of tone mapping operators. In *Proceedings of the 14th Pacific Conference on Computer Graphics and Applications*, pages 35– 44. National Taiwan University Press, 2006.

[83] Martin Čadík, Michael Wimmer, Laszlo Neumann, and Alessandro Artusi. Evaluation of HDR tone mapping methods using essential perceptual attributes. *Computers & Graphics*, 32:330–349, 2008.

[84] Tassio Castro, Alexandre Chapiro, Marcelo Cicconet, and Luiz Velho. Towards mobile hdr video. In Rafał Mantiuk and Erik Reinhard, editors, *Annex to the Conference Proceedings: Areas Papers on HDR*, Eurographics Area Papers on HDR, pages 57–62. Eurographics Association, April 2011.

[85] Edwin Earl Catmull. *A Subdivision Algorithm for Computer Display of Curved Surfaces*. PhD thesis, University of Utah, 1974.

[86] Alan Chalmers, Gerhard Bonnet, Francesco Banterle, Piotr Dubla, Kurt Debattista, Alessandro Artusi, and Christopher Moir. High-dynamic-range video solution. In *SIGGRAPH ASIA '09: ACM SIGGRAPH ASIA 2009 Art Gallery & Emerging Technologies: Adaptation*, pages 71–71, New York, NY, USA, 2009. ACM.

[87] Alan Chalmers, Brian Karr, Rossella Suma, and Kurt Debattista. Fifty shades of hdr. In *Digital Media Industry & Academic Forum (DMIAF)*. IEEE, July 2016.

[88] Abbas Chen, Tingand El Gamal. Optimal scheduling of capture times in a multiple capture imaging system. In *Proc. SPIE 4669, Sensors and Camera Systems for Scientific, Industrial, and Digital Photography Applications III*, volume 4669. SPIE, April 2002.

[89] Jiawen Chen, Sylvain Paris, and Frédo Durand. Real-time edge-aware image processing with the bilateral grid. *ACM Trans. Graph.*, 26(3):103, 2007.

[90] Kenneth Chiu, M. Herf, Peter Shirley, S. Swamy, Changyaw Wang, and Kurt Zimmerman. Spatially nonuniform scaling functions for high contrast images. In *Proceedings of Graphics Interface 93*, pages 245–253, San Francisco, CA, USA, May 1993. Morgan Kaufmann Publishers Inc.

[91] Hojin Cho, Seon Joo Kim, and Seungyong Lee. Single-shot high dynamic range imaging using coded electronic shutter. *Comput. Graph. Forum*, 33(7):329–338, October 2014.

[92] Prasun Choudhury and Jack Tumblin. The trilateral filter for high contrast images and meshes. In *EGRW '03: Proceedings of the 14th Eurographics Workshop on Rendering*, pages 186–196, Aire-la-Ville, Switzerland, 2003. Eurographics Association.

[93] CIE. Commission Internationale de l'Eclairage. Available at http://www.cie.co.at, December 2008.

[94] Petrik Clarberg and Tomas Akenine-Möller. Exploiting Visibility Correlation in Direct Illumination. *Computer Graphics Forum (Proceedings of EGSR 2008)*, 27(4):1125–1136, 2008.

[95] Petrik Clarberg and Tomas Akenine-Möller. Practical product importance sampling for direct illumination. *Computer Graphics Forum (Proceedings of Eurographics 2008)*, 27(2):681–690, 2008.

[96] Petrik Clarberg, Wojciech Jarosz, Tomas Akenine-Möller, and Henrik Wann Jensen. Wavelet importance sampling: Efficiently evaluating products of complex functions. *ACM Trans. Graph.*, 24(3):1166–1175, 2005.

[97] Jonathan Cohen, Chris Tchou, Tim Hawkins, and Paul Debevec. Real-Time High-Dynamic Range Texture Mapping. In *Eurographics Rendering Workshop*, June 2001.

[98] Massimiliano Corsini, Marco Callieri, and Paolo Cignoni. Stereo light probe. *Computer Graphics Forum*, 27(2):291–300, 2008.

[99] CypressSemiconductor. LUPA 1300-2. Available at http://www.cypress.com/, October 2016.

[100] Scott Daly. The visible differences predictor: An algorithm for the assessment of image fidelity. In *Digital Images and Human Vision*, pages 179–206, Cambridge, MA, USA, 1993. MIT Press.

[101] Scott Daly and Xiaofan Feng. Bit-depth extension using spatiotemporal microdither based on models of the equivalent input noise of the visual system. In *Proceedings of Color Imaging VIII: Processing, Hardcopy, and Applications*, pages 455–466, Bellingham, WA, USA, June 2003. SPIE.

[102] Scott Daly and Xiaofan Feng. Decontouring: Prevention and removal of false contour artifacts. In *Proceedings of Human Vision and Electronic Imaging IX*, pages 130–149, Bellingham, WA, USA, June 2004. SPIE.

[103] Herbert A. David. *The Method of Paired Comparisons, Second Edition*. Oxford University Press, Oxford, UK, 1988.

[104] Francesca De Simone, Giuseppe Valenzise, Paul Lauga, Frederic Dufaux, and Francesco Banterle. Dynamic range expansion of video sequences: a subjective quality assessment study. In *The 2nd IEEE Global Conference on Signal and Information Processing*, pages 1231–1235. IEEE, IEEE Signal Processing Society, December 2014.

[105] Kurt Debattista, Thomas Bashford-Rogers, and Alan Chalmers. Practical backwards compatible high dynamic range compression. In Alan Chalmers, Patrizio Campisi, Peter Shirley, and Igor Olaziola, editors, *High Dynamic Range Video: Concepts, Technologies and Applications*, chapter 11. Elsevier, 2016.

[106] Kurt Debattista, Thomas Bashford-Rogers, Elmedin Selmanović, Ratnajit Mukherjee, and Alan Chalmers. Optimal exposure compression for high dynamic range content. *The Visual Computer*, 31(6-8):1089–1099, 2015.

[107] Paul Debevec. Rendering synthetic objects into real scenes: Bridging traditional and image-based graphics with global illumination and high dynamic range photography. In *SIGGRAPH '98: Proceedings of the 25th Annual Conference on Computer Graphics and Interactive Techniques*, pages 189–198, New York, NY, USA, 1998. ACM.

[108] Paul Debevec. A median cut algorithm for light probe sampling. In *SIGGRAPH '05: ACM SIGGRAPH 2005 Posters*, page 66, New York, NY, USA, 2005. ACM.

[109] Paul Debevec. Virtual cinematography: Relighting through computation. *Computer*, 39(8):57–65, 2006.

[110] Paul Debevec, Tim Hawkins, Chris Tchou, Haarm-Pieter Duiker, Westley Sarokin, and Mark Sagar. Acquiring the reflectance field of a human face. In *SIGGRAPH '00: Proceedings of the 27th Annual Conference on Computer Graphics and Interactive Techniques*, pages 145–156, New York, NY, USA, 2000. ACM Press/Addison-Wesley Publishing Co.

[111] Paul Debevec and Jitendra Malik. Recovering high dynamic range radiance maps from photographs. In *SIGGRAPH '97: Proceedings of the 24th Annual Conference on Computer Graphics and Interactive Techniques*, pages 369–378, New York, NY, USA, 1997. ACM Press/Addison-Wesley Publishing Co.

[112] Paul Debevec and Erik Reinhard. High dynamic range imaging: Theory and applications. In *ACM SIGGRAPH 2006 Courses*, New York, NY, USA, 2006. ACM.

[113] Piotr Didyk, Rafał Mantiuk, Matthias Hein, and Hans-Peter Seidel. Enhancement of bright video features for HDR displays. *Computer Graphics Forum*, 27(4):1265–1274, 2008.

[114] Yue Dong, Xin Tong, Fabio Pellacini, and Baining Guo. Printing Spatially-varying Reflectance for Reproducing HDR Images. *ACM Trans. Graph.*, 31(4):40:1–40:7, July 2012.

[115] Frederic Drago, William Martens, Karol Myszkowski, and Hans-Peter Seidel. Perceptual evaluation of tone mapping operators with regard to similarity and preference. Research Report MPI-I-2002-4-002, Max-Planck-Institut für Informatik, Stuhlsatzenhausweg 85, 66123 Saarbrücken, Germany, August 2002.

[116] Frederic Drago, Karol Myszkowski, Thomas Annen, and Norishige Chiba. Adaptive logarithmic mapping for displaying high contrast scenes. *Computer Graphics Forum*, 22(3):419–426, 2003.

[117] Frédo Durand and Julie Dorsey. Interactive tone mapping. In *Proceedings of the Eurographics Workshop on Rendering Techniques 2000*, pages 219–230, London, UK, 2000. Springer-Verlag.

[118] Frédo Durand and Julie Dorsey. Fast bilateral filtering for the display of high-dynamic-range images. *ACM Trans. Graph.*, 21(3):257–266, 2002.

[119] David E. Jacobs, Orazio Gallo, Emily A. Cooper, Kari Pulli, and Marc Levoy. Simulating the visual experience of very bright and very dark scenes. *ACM Trans. Graph.*, 34(3):25:1–25:15, May 2015.

[120] Fritz Ebner and Mark D. Fairchild. Development and testing of a color space (IPT) with improved hue uniformity. In *6th Color and Imaging Conference, CIC 1998, November 17–20*, pages 8–13. IS&T - The Society for Imaging Science and Technology, 1998.

[121] Gabriel Eilertsen, Rafał K. Mantiuk, and Jonas Unger. Real-time noise-aware tone mapping. *ACM Trans. Graph.*, 34(6):198:1–198:15, October 2015.

[122] Gabriel Eilertsen, Rafał K. Mantiuk, and Jonas Unger. A high dynamic range video codec optimized by large-scale testing. In *International Conference on Image Processing (ICIP)*, pages 1379–1383. IEEE Computer Society, September 2016.

[123] Gabriel Eilertsen, Robert Wanat, Rafał K. Mantiuk, and Jonas Unger. Evaluation of tone mapping operators for hdr-video. *Computer Graphics Forum*, 32(7):275–284, 2013.

[124] Mark D. Fairchild. *Color Appearance Models*. Wiley, 3rd edition, 2013.

[125] Mark D. Fairchild and Garrett M. Johnson. Meet icam: A next-generation color appearance model. In *The Tenth Color Imaging Conference*, pages 33–38. IS&T - The Society for Imaging Science and Technology, 2002.

[126] Zeev Farbman, Raanan Fattal, Dani Lischinski, and Richard Szeliski. Edge-preserving decompositions for multi-scale tone and detail manipulation. *ACM Trans. Graph.*, 27(3):67:1–67:10, August 2008.

[127] Hany Farid. Blind inverse gamma correction. *IEEE Transactions on Image Processing*, 10(10):1428–1433, October 2001.

[128] Raanan Fattal. Edge-avoiding wavelets and their applications. *ACM Trans. Graph.*, 28(3):22:1–22:10, July 2009.

[129] Raanan Fattal, Dani Lischinski, and Michael Werman. Gradient domain high dynamic range compression. *ACM Trans. Graph.*, 21(3):249–256, 2002.

[130] James A. Ferwerda, Sumanta N. Pattanaik, Peter Shirley, and Donald P. Greenberg. A model of visual adaptation for realistic image synthesis. In *SIGGRAPH '96: Proceedings of the 23rd Annual Conference on Computer Graphics and Interactive Techniques*, pages 249–258, New York, NY, USA, 1996. ACM.

[131] James A. Ferwerda, Peter Shirley, Sumanta N. Pattanaik, and Donald P. Greenberg. A model of visual masking for computer graphics. In *SIGGRAPH '97: Proceedings of the 24th Annual Conference on Computer Graphics and Interactive Techniques*, pages 143–152, New York, NY, USA, 1997. ACM Press/Addison-Wesley Publishing Co.

[132] Ezio Filippi. *Geografia della pietra di Prun*. Gutemberg, Povegliano Veronese, Italy, 1982.

[133] International Organization for Standardization (ISO). ISO 2720:1974: Photography – General purpose photographic exposure meters (photoelectric type) – Guide to product specification, 1974.

[134] International Organization for Standardization (ISO). ISO/IEC 15444-1:2004 Information technology – JPEG 2000 image coding system: Core coding system – Part 1, 2004.

[135] International Organization for Standardization (ISO). ISO/IEC 29199-2:2010: Information technology – JPEG XR image coding system – Part 2: Image coding specification, 2010.

[136] International Organization for Standardization (ISO). ISO/IEC 15444-10:2011: Information technology – JPEG 2000 image coding system: Extensions for three-dimensional data, November 2011.

[137] Jan Froehlich, Stefan Grandinetti, Bernd Eberhardt, Simon Walter, Andreas Schilling, and Harald Brendel. Creating cinematic wide gamut hdr-video for the evaluation of tone mapping operators and hdr-displays. In *Proceedings of SPIE Electronic Imaging*, volume 9023, March 2014.

[138] Orazio Gallo, Marius Tico, Roberto Manduchi, Natasha Gelfand, and Kari Pulli. Metering for exposure stacks. *Computer Graphics Forum (Proceedings of Eurographics)*, 31:479–488, 2012.

[139] Jens-Uwe Garbas and Herbert Thoma. Temporally coherent luminance-to-luma mapping for high dynamic range video coding with h. 264/avc. In *IEEE International Conference on Acoustics, Speech and Signal Processing (ICASSP)*, pages 829–832. IEEE Computer Society, 2011.

[140] Eduardo S. L. Gastal and Manuel M. Oliveira. Adaptive manifolds for real-time high-dimensional filtering. *ACM Trans. Graph.*, 31(4):33:1–33:13, July 2012.

[141] Abhijeet Ghosh, Arnaud Doucet, and Wolfgang Heidrich. Sequential sampling for dynamic environment maps. In *SIGGRAPH '06: ACM SIGGRAPH 2006 Sketches*, page 157, New York, NY, USA, 2006. ACM.

[142] Alan Gilchrist, Christos Kossyfidis, Frederick Bonato, Tiziano Agostini, Xiaojun Li Joseph Cataliotti, Branka Spehar, Vidal Annan, and Elias Economou. An anchoring theory of lightness perception. *Psychological Review*, 106(4):795–834., October 1999.

[143] Michael Goesele, Wolfgang Heidrich, and Hans-Peter Seidel. Color calibrated high dynamic range imaging with icc profiles. In *Proceedings of the IS&T/SID 9th Color Imaging Conference*, pages 286–290. Society for Imaging Science and Technology, November 2001.

[144] Rafael C. Gonzalez and Richard E. Woods. *Digital Image Processing*. Addison-Wesley Longman Publishing Co., Inc., Boston, MA, USA, 2001.

[145] Nolan Goodnight, Rui Wang, Cliff Woolley, and Greg Humphreys. Interactive time-dependent tone mapping using programmable graphics hardware. In *Proceedings Eurographics Symposium on Rendering*, pages 26–37. Eurographics Association, 2003.

[146] A. Ardeshir Goshtasby. Fusion of multi-exposure images. *Image Vision Comput.*, 23(6):611–618, June 2005.

[147] Miguel Granados, Boris Ajdin, Michael Wand, Christian Theobalt, Hans-Peter Seidel, and Hendrik P. A. Lensch. Optimal hdr reconstruction with linear digital cameras. In *CVPR*, pages 215–222, 2010.

[148] Computer Graphics and Image Processing Visual Computing Laboratory. Liu hdrv repository. Available at http://hdrv.org, January 2017.

[149] GrassValley. Thomsongrassvalley. Available at http://www.grassvalley. com/, October 2016.

[150] Robin Green. Spherical harmonics lighting: The gritty details. In *Game Developers Conference*, pages 1–47, 2003.

[151] Ned Greene. Environment mapping and other applications of world projections. *IEEE Computer Graphics and Applications*, 6(11):21–29, November 1986.

[152] Michael D. Grossberg and Shree K. Nayar. What can be Known about the Radiometric Response Function from Images. In *European Conference on Computer Vision (ECCV)*, volume IV, pages 189–205. Springer, May 2002.

[153] Michael D. Grossberg and Shree K. Nayar. Modeling the Space of Camera Response Functions. *IEEE Transactions on Pattern Analysis and Machine Intelligence*, 26(10):1272–1282, October 2004.

[154] Stanford Graphics Group. The Stanford 3D scanning repository. Available at http://graphics.stanford.edu/data/3Dscanrep/, December 2008.

[155] Yulia Gryaditskaya, Tania Pouli, Erik Reinhard, Karol Myszkowski, and Hans-Peter Seidel. Motion aware exposure bracketing for hdr video. In *Proceedings of the 26th Eurographics Symposium on Rendering*, EGSR '15, pages 119–130, Aire-la-Ville, Switzerland, 2015. Eurographics Association.

[156] Yulia Gryaditskya, Tania Pouli, Erik Reinhard, and Hans-Peter Seidel. Sky based light metering for high dynamic range images. *Computer Graphics Forum*, 33(7):61–69, October 2014.

[157] J. Gu, Y. Hitomi, T. Mitsunaga, and S.K. Nayar. Coded Rolling Shutter Photography: Flexible Space-Time Sampling. In *Proceeding of the International Conference on Computational Photography (ICCP)*, pages 248–260, Washington, DC, USA, March 2010. IEEE Computer Society.

[158] Dong Guo, Yuan Cheng, Shaojie Zhuo, and Terence Sim. Correcting overexposure in photographs. In *The Twenty-Third IEEE Conference on Computer Vision and Pattern Recognition, CVPR, 13–18 June*, pages 515–521, 2010.

[159] Toshiya Hachisuka and Henrik Wann Jensen. Stochastic progressive photon mapping. *ACM Trans. Graph.*, 28(5):141:1–141:8, December 2009.

[160] Yoav HaCohen, Eli Shechtman, Dan B. Goldman, and Dani Lischinski. Non-rigid dense correspondence with applications for image enhancement. *ACM Trans. Graph.*, 30(4):70:1–70:10, July 2011.

[161] Saghi Hajisharif, Jonas Unger, and Joel Kronander. HDR reconstruction for alternating gain (ISO) sensor readout. In *Proceedings of Eurographics Short Papers*, May 2014.

[162] Philippe Hanhart, Martin Řeřábek, and Touradj Ebrahimi. Towards high dynamic range extensions of hevc: subjective evaluation of potential coding technologies. In *SPIE Optical Engineering+ Applications*, pages 95990G–95990G. International Society for Optics and Photonics, 2015.

[163] Ami Harten, Bjorn Engquist, Stanley Osher, and Sukumar R. Chakravarthy. Uniformly high order accurate essentially non-oscillatory schemes, 111. *J. Comput. Phys.*, 71(2):231–303, August 1987.

[164] Samuel W. Hasinoff, Frédo Durand, and William T. Freeman. Noise-optimal capture for high dynamic range photography. In *The Twenty-Third IEEE Conference on Computer Vision and Pattern Recognition, CVPR 2010, San Francisco, CA, USA, 13-18 June 2010*, pages 553–560, 2010.

[165] Jonathan Hatchett, Kurt Debattista, Ratnajit Mukherjee, Thomas Bashford-Rogers, and Alan Chalmers. An evaluation of power transfer functions for hdr video compression. *The Visual Computer*, 2016.

[166] Vlastimil Havran, Kirill Dmitriev, and Hans-Peter Seidel. Goniometric Diagram Mapping for Hemisphere. In *Eurographics 2003 - Short Presentations*, pages 293–300. Eurographics Association, 2003.

[167] Vlastimil Havran, Miloslaw Smyk, Grzegorz Krawczyk, Karol Myszkowski, and Hans-Peter Seidel. Interactive system for dynamic scene lighting using captured video environment maps. In *Proceedings of the Sixteenth Eurographics Conference on Rendering Techniques*, EGSR '05, pages 31–42, Aire-la-Ville, Switzerland, 2005. Eurographics Association.

[168] Mary M. Hayhoe, Norma I. Benimoff, and D. C. Hood. The time course of multiplicative and subtractive adaptation process. *Vision Research*, 27:1981–1996, 1987.

[169] Kaiming He, Jian Sun, and Xiaoou Tang. Guided image filtering. In *Proceedings of the 11th European Conference on Computer Vision: Part I*, ECCV'10, pages 1–14, Berlin, Heidelberg, 2010. Springer-Verlag.

[170] Donald Healy and O. Mitchell. Digital video bandwidth compression using block truncation coding. *IEEE Transactions on Communications*, 29(12):1809–1817, December 1981.

[171] Berthold K. Horn. Determining lightness from an image. *Computer Graphics and Image Processing*, 3(1):277–299, December 1974.

[172] Berthold K. P. Horn. *Robot Vision*. MIT Press, Cambridge, MA, USA, 1986.

[173] David Hough. Applications of the proposed IEEE-754 standard for floating point arithmetic. *Computer*, 14(3):70–74, March 1981.

[174] Jun Hu, Orazio Gallo, and Kari Pulli. Exposure stacks of live scenes with hand-held cameras. In *Computer Vision - ECCV 2012 - 12th European Conference on Computer Vision, October 7–13, Proceedings, Part I*, pages 499–512, 2012.

[175] Jun Hu, Orazio Gallo, Kari Pulli, and Xiaobai Sun. HDR deghosting: How to deal with saturation? In *2013 IEEE Conference on Computer Vision and Pattern Recognition, June 23–28*, pages 1163–1170, 2013.

[176] Vedad Hulusić, Kurt Debattista, Giuseppe Valenzise, and Frédéric Dufaux. A model of perceived dynamic range for hdr images. *Signal Processing: Image Communication*, 2016.

[177] Vedad Hulusić, Giuseppe Valenzise, Edoardo Provenzi, Kurt Debattista, and Frederic Dufaux. Perceived dynamic range of hdr images. In *International Conference on Quality of Multimedia Experience (QoMEX)*, pages 1–6. IEEE Computer Society, 2016.

[178] Rober W. G. Hunt. *The Reproduction of Colour*. Fountain Press Ltd, Kingston-upon-Thames, England, 1995.

[179] Yongqing Huo, Fan Yang, Le Dong, and Vincent Brost. Physiological inverse tone mapping based on retina response. *The Visual Computer*, 30:507–517, May 2014.

[180] Industrial Light & Magic. OpenEXR. Available at https://github.com/openexr/openexr, October 2016.

[181] Konstantine Iourcha, Krishna Nayak, and Zhou Hong. System and method for fixed-rate block-based image compression with inferred pixel values. Patent no. 5,956,431, 1997.

[182] Piti Irawan, James A. Ferwerda, and Stephen R. Marschner. Perceptually based tone mapping of high dynamic range image streams. In Oliver Deussen, Alexander Keller, Kavita Bala, Philip Dutr, Dieter W. Fellner, and Stephen N. Spencer, editors, *Proceedings of the 16th Eurographics Symposium on Rendering*, pages 231–242. Eurographics Association, 2005.

[183] ITU. ITU-R BT.709, basic parameter values for the HDTV standard for the studio and for international programme exchange. In *Standard Recommendation 709, International Telecommunication Union*, 1990.

[184] ITU-R. Proposed preliminary draft new Report - Image dynamic range in television systems. ITU-R WP6C Contribution 146, April 2013.

[185] Kei Iwasaki, Yoshinori Dobashi, Fujiichi Yoshimoto, and Tomoyuki Nishita. Precomputed radiance transfer for dynamic scenes taking into account light interreflection. In *Rendering Techniques 2007: 18th Eurographics Workshop on Rendering*, pages 35–44, Aire-la-Ville, Switzerland, June 2007. Eurographics Association.

[186] James T. Kajiya. The rendering equation. *SIGGRAPH Computer Graphics*, 20(4):143–150, 1986.

[187] Nima Khademi Kalantari, Eli Shechtman, Connelly Barnes, Soheil Darabi, Dan B. Goldman, and Pradeep Sen. Patch-based high dynamic range video. *ACM Trans. Graph.*, 32(6):202:1–202:8, November 2013.

[188] Sing Bing Kang, Matthew Uyttendaele, Simon Winder, and Richard Szeliski. High dynamic range video. *ACM Trans. Graph.*, 22(3):319–325, 2003.

[189] Levent Karacan, Erkut Erdem, and Aykut Erdem. Structure-preserving image smoothing via region covariances. *ACM Trans. Graph.*, 32(6):176:1–176:11, November 2013.

[190] B. Karr, A.G. Chalmers, and K. Debattista. High dynamic range digital imaging of spacecraft. In F. Dufaux, P. Le Callet, R. Mantiuk, and M. Mrak, editors, *High Dynamic Range Videos - From Acquisition to Display and Applications*, chapter 20, pages 519–547. Elsevier, April 2016.

[191] Petr Kellnhofer, Tobias Ritschel, Karol Myszkowski, Elmar Eisemann, and Hans-Peter Seidel. Modeling luminance perception at absolute threshold. *Computer Graphics Forum (Proc. of EGSR)*, 34(4):1–10, June 2015.

[192] Maurice Kendall. *Rank Correlation Methods, Fourth Edition*. Griffin Ltd., Baltimore, MD, USA, 1975.

[193] Min H. Kim and Jan Kautz. Characterization for High Dynamic Range Imaging. *Computer Graphics Forum*, 27(2):691–697, April 2008.

[194] Min H. Kim, Tim Weyrich, and Jan Kautz. Modeling human color perception under extended luminance levels. *ACM Trans. Graph.*, 28(3):27:1–27:9, July 2009.

[195] Adam G. Kirk and James F. O'Brien. Perceptually based tone mapping for low-light conditions. *ACM Trans. Graph.*, 30(4):42:1–42:10, July 2011.

[196] Chris Kiser, Erik Reinhard, Mike Tocci, and Nora Tocci. Real time automated tone mapping system for hdr video. In *In Proc. of the IEEE Int. Conference on Image Processing*. IEEE Computer Society, 2012.

[197] Gunter Knittel, Andreas Schilling, Anders Kugler, and Wolfgang Strasser. Hardware for superior texture performance. *Computers & Graphics*, 20(4):475–481, 1996.

[198] Thomas Kollig and Alexander Keller. Efficient illumination by high dynamic range images. In *EGRW '03: Proceedings of the 14th Eurographics Workshop on Rendering*, pages 45–50, Aire-la-Ville, Switzerland, 2003. Eurographics Association.

[199] Johannes Kopf, Michael F. Cohen, Dani Lischinski, and Matt Uyttendaele. Joint bilateral upsampling. *ACM Trans. Graph.*, 26(3):96, 2007.

[200] Pavel Korshunov, Philippe Hanhart, Thomas Richter, Alessandro Artusi, Rafał Mantiuk, and Touradj Ebrahimi. Subjective quality assessment database of HDR images compressed with JPEG XT. In *7th International Workshop on Quality of Multimedia Experience (QoMEX)*, 2015.

[201] Rafael Pacheco Kovaleski and Manuel M. Oliveira. High-quality brightness enhancement functions for real-time reverse tone mapping. *Vis. Comput.*, 25(5–7):539–547, 2009.

[202] Rafael Pacheco Kovaleski and Manuel M. Oliveira. High-quality reverse tone mapping for a wide range of exposures. In *27th SIBGRAPI Conference on Graphics, Patterns and Images*, pages 49–56. IEEE Computer Society, August 2014.

[203] Alper Koz and Frederic Dufaux. A comparative survey on high dynamic range video compression. In *SPIE Optical Engineering+ Applications*, pages 84990E–84990E. International Society for Optics and Photonics, 2012.

[204] Grzegorz Krawczyk, Karol Myszkowski, and Hans-Peter Seidel. Lightness perception in tone reproduction for high dynamic range images. *Computer Graphics Forum (Proceeding of Eurographics 2005)*, 24(3):635–645, 2005.

[205] Grzegorz Krawczyk, Karol Myszkowski, and Hans-Peter Seidel. Contrast restoration by adaptive countershading. *Computer Graphics Forum (Proceeding of Eurographics 2007)*, 26(3):581–590, 2007.

[206] Joel Kronander, Stefan Gustavson, Gerhard Bonnet, Anders Ynnerman, and Jonas Unger. A unified framework for multi-sensor HDR video reconstruction. *Signal Processing: Image Communications*, 29(2):203–215, 2014.

[207] Jiangtao Kuang, Garrett M. Johnson, and Mark D. Fairchild. icam06: A refined image appearance model for HDR image rendering. *J. Vis. Comun. Image Represent.*, 18(5):406–414, 2007.

[208] Jiangtao Kuang, Hiroshi Yamaguchi, Garrett M. Johnson, and Mark D. Fairchild. Testing HDR image rendering algorithms. In *IS&T/SID 12th Color Imaging Conference*, pages 315–320, Scottsdale, AZ, USA, 2004. SPIE.

[209] Jiangtao Kuang, Hiroshi Yamaguchi, Changmeng Liu, Garrett M. Johnson, and Mark D. Fairchild. Evaluating HDR rendering algorithms. *ACM Transaction on Applied Perception*, 4(2):9, 2007.

[210] Pin-Hung Kuo, Huai-Jen Liang, Chi-Sun Tang, and Shao-Yi Chien. Automatic high dynamic range hallucination in inverse tone mapping. In *IEEE 16th International Workshop on Multimedia Signal Processing (MMSP), September 22–24*, pages 1–6, 2014.

[211] Eric P. Lafortune and Yves D. Willems. Bi-Directional Path Tracing. In *Proceedings of the Third International Conference on Computational Graphics and Visualization Techniques (Compugraphics '93)*, pages 145–153, December 1993.

[212] Edwin Land. Recent advances in retinex theory. *Vision Research*, 19(1):7–21, 1986.

[213] Hayden Landis. Production-ready global illumination. In *SIGGRAPH Course Notes* 16, pages 87–101, 2002.

[214] Gregory Ward Larson. Logluv encoding for full-gamut, high-dynamic range images. *Journal of Graphics Tools*, 3(1):15–31, 1998.

[215] Gregory Ward Larson, Holly Rushmeier, and Christine Piatko. A visibility matching tone reproduction operator for high dynamic range scenes. *IEEE Transactions on Visualization and Computer Graphics*, 3(4):291–306, 1997.

[216] Patrick Ledda, Alan Chalmers, Tom Troscianko, and Helge Seetzen. Evaluation of tone mapping operators using a high dynamic range display. In *SIGGRAPH '05: ACM SIGGRAPH 2005 Papers*, pages 640–648, New York, NY, USA, 2005. ACM.

[217] Patrick Ledda, Luis Paulo Santos, and Alan Chalmers. A local model of eye adaptation for high dynamic range images. In *AFRIGRAPH '04: Proceedings of the 3rd International Conference on Computer Graphics, Virtual Reality, Visualisation and Interaction in Africa*, pages 151–160, New York, NY, USA, 2004. ACM.

[218] Patrick Ledda, Greg Ward, and Alan Chalmers. A wide field, high dynamic range, stereographic viewer. In *GRAPHITE '03: Proceedings of the 1st International Conference on Computer Graphics and Interactive Techniques in Australasia and South East Asia*, pages 237–244, New York, NY, USA, 2003. ACM.

[219] Chu Lee and Chang-Su Kim. Gradient domain tone mapping of high dynamic range videos. In *IEEE International Conference on Image Processing*, pages 461–464. IEEE Computer Society, November 2007.

[220] Chu Lee and Chang-Su Kim. Rate-distortion optimized compression of high dynamic range videos. In *16th European Signal Processing Conference (EUSIPCO 2008)*, pages 461–464, 2008.

[221] Joon-Young Lee, Boxin Shi, Y. Matsushita, In-So Kweon, and K. Ikeuchi. Radiometric calibration by transform invariant low-rank structure. In *Proceedings of the 2011 IEEE Conference on Computer Vision and Pattern Recognition*, CVPR '11, pages 2337–2344, Washington, DC, USA, 2011. IEEE Computer Society.

[222] Anat Levin, Dani Lischinski, and Yair Weiss. Colorization using optimization. *ACM Trans. Graph.*, 23(3):689–694, 2004.

[223] Marc Levoy and Pat Hanrahan. Light field rendering. In *SIGGRAPH '96: Proceedings of the 23rd Annual Conference on Computer Graphics and Interactive Techniques*, pages 31–42, New York, NY, USA, 1996. ACM.

[224] Shigang Li. Real-time spherical stereo. In *Proceedings of the 18th International Conference on Pattern Recognition (ICPR '06)*, pages 1046–1049. IEEE Computer Society, 2006.

[225] Yuanzhen Li, Lavanya Sharan, and Edward H. Adelson. Compressing and companding high dynamic range images with subband architectures. *ACM Trans. Graph.*, 24(3):836–844, 2005.

[226] Huei-Yung Lin and Wei-Zhe Chang. High dynamic range imaging for stereoscopic scene representation. In *Proceedings of the International Conference on Image Processing (ICIP), 7–10 November*, pages 4305–4308, 2009.

[227] Stephen Lin, Jinwei Gu, Shuntaro Yamazaki, and Heung-Yeung Shum. Radiometric calibration from a single image. In *CVPR 2004: Proceedings of the 2004 IEEE Conference on Computer Vision and Pattern Recognition (CVPR2004)*, pages 938–945, Washington, DC, USA, 2004. IEEE Computer Society.

[228] Stephen Lin and Lei Zhang. Determining the radiometric response function from a single grayscale image. In *CVPR '05: Proceedings of the 2005 IEEE Computer Society Conference on Computer Vision and Pattern Recognition (CVPR'05)*, pages 66–73, Washington, DC, USA, 2005. IEEE Computer Society.

[229] Dani Lischinski, Zeev Farbman, Matt Uyttendaele, and Richard Szeliski. Interactive local adjustment of tonal values. *ACM Trans. Graph.*, 25(3):646–653, 2006.

[230] Jing Liu, Nikolce Stefanoski, Tunç Ozan Aydın, Anselm Grundhöfer, and Aljosa Smolic. Chromatic calibration of an HDR display using 3d octree forests. In *International Conference on Image Processing (ICIP), September 27–30*, pages 4317–4321, 2015.

[231] Xiaopei Liu, Liang Wan, Yingge Qu, Tien-Tsin Wong, Stephen Lin, Chi-Sing Leung, and Pheng-Ann Heng. Intrinsic colorization. *ACM Trans. Graph.*, 27(5):1–9, 2008.

[232] Stuart P. Lloyd. Least squares quantization in pcm. *IEEE Transactions on Information Theory*, 28(2):129–137, March 1982.

[233] Eva Lübbe. *Colours in the Mind-Colour Systems in Reality: A formula for Colour Saturation*. Books on Demand GmbH, 2008.

[234] Jeffrey Lubin. *A Visual Discrimination Model for Imaging System Design and Evaluation*, pages 245–283. World Scientific Publishers, River Edge, NJ, USA, 1995.

[235] Thomas Luft, Carsten Colditz, and Oliver Deussen. Image enhancement by unsharp masking the depth buffer. *ACM Trans. Graph.*, 25(3):1206–1213, 2006.

[236] Ajay Luthra, Edouard Francois, and Walt Husak. Call for evidence (cfe) for hdr and wcg video coding. *ISO/IEC JTC1/SC29/WG11 MPEG2014 N*, 15083, 2015.

[237] Zicong Mai, Hassan Mansour, Rafał Mantiuk, Panos Nasiopoulos, Rabab Ward, and Wolfgang Heidrich. Optimizing a tone curve for backward-compatible high dynamic range image and video compression. *IEEE Transactions on Image Processing*, 20(6):1558–1571, 2011.

[238] Stephen Mangiat and Jerry D. Gibson. Spatially adaptive filtering for registration artifact removal in HDR video. In *18th IEEE International Conference on Image Processing (ICIP), September 11–14*, pages 1317–1320, 2011.

[239] Steve Mann and Rosalind W. Picard. Being 'undigital' with digital cameras: Extending dynamic range by combining differently exposed pictures. In *Proceedings of IS&T 48th Annual Conference*, pages 422–428. Society for Imaging Science and Technology, May 1995.

[240] Radosław Mantiuk and Mateusz Markowski. Gaze-dependent tone mapping. *Lecture Notes in Computer Science (Proc. of ICIAR'13 Conference)*, 7950:426–433, 2013.

[241] Radosław Mantiuk, theł Mantiuk, Anna Tomaszweska, and Wolfgang Heidrich. Color correction for tone mapping. *Computer Graphics Forum*, 28(2):193–202, 2009.

[242] Rafał Mantiuk, Scott Daly, and Louis Kerofsky. Display adaptive tone mapping. *ACM Trans. Graph.*, 27(3):1–10, 2008.

[243] Rafał Mantiuk, Scott Daly, Karol Myszkowski, and Hans-Peter Seidel. Predicting visible differences in high dynamic range images—model and its calibration. In Bernice E. Rogowitz, Thrasyvoulos N. Pappas, and Scott J. Daly, editors, *Human Vision and Electronic Imaging X, IST SPIE's 17th Annual Symposium on Electronic Imaging*, pages 204–214. SPIE, 2005.

[244] Rafał Mantiuk, Alexander Efremov, Karol Myszkowski, and Hans-Peter Seidel. Backward compatible high dynamic range mpeg video compression. *ACM Trans. Graph.*, 25(3):713–723, 2006.

[245] Rafał Mantiuk, Grzegorz Krawczyk, Karol Myszkowski, and Hans-Peter Seidel. Perception-motivated high dynamic range video encoding. *ACM Trans. Graph.*, 23(3):733–741, 2004.

[246] Rafał Mantiuk, Karol Myszkowski, and Hans-Peter Seidel. Visible difference predicator for high dynamic range images. In *Proceedings of IEEE International Conference on Systems, Man and Cybernetics*, pages 2763–2769, 2004.

[247] Rafał Mantiuk and Hans-Peter Seidel. Modeling a generic tone-mapping operator. *Computer Graphics Forum*, 27(2):699–708, April 2008.

[248] Rafał K. Mantiuk, Kil Joong Kim, Allan G. Rempel, and Wolfgang Heidrich. Hdr-vdp-2: A calibrated visual metric for visibility and quality predictions in all luminance conditions. *ACM Trans. Graph.*, 30(4):40:1–40:14, July 2011.

[249] Rafał K. Mantiuk, Thomas Richter, and Alessandro Artusi. Fine-tuning JPEG-XT compression performance using large-scale objective quality testing. In *Proceedings of the IEEE international Conference on image Processing (ICIP)*, pages 2152–2156. IEEE Computer Society, September 2016.

[250] Rafał K. Mantiuk, Anna Tomaszewska, and Radosław Mantiuk. Comparison of four subjective methods for image quality assessment. *Computer Graphics Forum*, 31(8):2478–2491, 2012.

[251] Belen Masia, Sandra Agustin, Roland W. Fleming, Olga Sorkine, and Diego Gutierrez. Evaluation of reverse tone mapping through varying exposure conditions. *ACM Trans. Graph.*, 28(5):1–8, 2009.

[252] Belen Masia, Ana Serrano, and Diego Gutierrez. Dynamic range expansion based on image statistics. *Multimedia Tools and Applications*, pages 1–18, 2015.

[253] S.Z. Masood, J. Zhu, and M.F. Tappen. Automatic correction of saturated regions in photographs using cross-channel correlation. *Computer Graphics Forum*, 28:1861–1869, October 2009.

[254] George Mather. *Foundations of Perception, First Edition*. Psychology Press, Hove, East Sussex, UK, March 2006.

[255] Mathworks. Image Processing Toolbox. Available at http://www.mathworks.com/products/image/, July 2010.

[256] John J. McCann and Alessandro Rizzi. Veiling glare: the dynamic range limit of HDR images. In *Human Vision and Electronic Imaging XII, January 29 – February 1*, page 649213, 2007.

[257] Miguel Melo, Maximinho Bessa, Kurt Debattista, and Alan Chalmers. Evaluation of tone-mapping operators for hdr video under different ambient luminance levels. *Computer Graphics Forum*, 34(8):38–49, 2015.

[258] Miguel Melo, Maximino Bessa, Kurt Debattista, and Alan Chalmers. Evaluation of hdr video tone mapping for mobile devices. *Signal Processing: Image Communication*, 29(2):247–256, 2014.

[259] Jerry M. Mendel. Tutorial on higher-order statistics (spectra) in signal processing and system theory: Theoretical results and some applications. *Proceedings of the IEEE*, 79(3):278–305, March 1991.

[260] T. Mertens, J. Kautz, and F. Van Reeth. Exposure Fusion: A Simple and Practical Alternative to High Dynamic Range Photography. *Computer Graphics Forum*, 28(1):161–171, March 2009.

[261] Tom Mertens, Jan Kautz, and Frank Van Reeth. Exposure fusion. In *PG '07: Proceedings of the 15th Pacific Conference on Computer Graphics and Applications*, pages 382–390, Washington, DC, USA, 2007. IEEE Computer Society.

[262] Laurence Meylan, Scott Daly, and Sabine Süsstrunk. The Reproduction of Specular Highlights on High Dynamic Range Displays. In *IST/SID 14th Color Imaging Conference*, pages 333–338, Scottsdale, AZ, USA, 2006.

[263] Laurence Meylan, Scott Daly, and Sabine Süsstrunk. Tone Mapping for High Dynamic Range Displays. In *Proc. IS&T/SPIE Electronic Imaging: Human Vision and Electronic Imaging XII*. SPIE, 2007.

[264] Laurence Meylan and Sabine Süsstrunk. High dynamic range image rendering with a retinex-based adaptive filter. *IEEE Transactions on Image Processing*, 15(9):2820–2830, 2006.

[265] Ehsan Miandji, Joel Kronander, and Jonas Unger. Compressive Image Reconstruction in Reduced Union of Subspaces. *Comput. Graph. Forum*, 34(2):33–44, 2015.

[266] Microsoft. BC6H Format. Available at https://msdn.microsoft.com/en-us/library/windows/desktop/hh308952(v=vs.85).aspx#bc6h_implementation, 2010.

[267] Microsoft. DXGI Format. Available at https://msdn.microsoft.com/en-us/library/windows/desktop/bb173059(v=vs.85).aspx, October 2016.

[268] Microsoft Corporation. Directx. Available at http://msdn.microsoft.com/en-us/directx/default.aspx, February 2010.

[269] Gene Miller and C. Robert Hoffman. Illumination and reflection maps: Simulated objects in simulated and real environments. In *Siggraph '84 Advanced Computer Graphics Animation Seminar Note*, New York, NY, USA, July 1984. ACM Press.

[270] Scott Miller, Mahdi Nezamabadi, and Scott Daly. Perceptual signal coding for more efficient usage of bit codes. *SMPTE Motion Imaging Journal*, 122(4):52–59, 2013.

[271] Tomoo Mitsunaga and Shree K. Nayar. Radiometric self calibration. *IEEE Computer Society Conference on Computer Vision and Pattern Recognition (CVPR)*, 1:1374, 1999.

[272] Nathan Moroney, Mark D. Fairchild, Robert W. G. Hunt, Changjun Li, M. Ronnier Luo, and Todd Newman. The ciecam02 color appearance model. In *IS&T/SID 10th Color Imaging Conference*, pages 23–27, 2002.

[273] Jan Morovic. *Color Gamut Mapping*. John Wiley & Sons Ltd., 2008.

[274] Ajit Motra and Herbert Thoma. An adaptive logluv transform for high dynamic range video compression. In *Proceedings of the International Conference on Image Processing (ICIP)*, pages 2061–2064, September 2010.

[275] Moving Picture Experts Group. Iso/iec 14496-2: 1999 (MPEG-4, Part 2). In *ISO*, 1999.

[276] Ratnajit Mukherjee, Kurt Debattista, Thomas Bashford-Rogers, Peter Vangorp, Rafał K. Mantiuk, Maximino Bessa, Brian Waterfield, and Alan Chalmers. Objective and subjective evaluation of high dynamic range video compression. *Signal Processing: Image Communication*, 47:426–437, 2016.

[277] Jacob Munkberg, Petrik Clarberg, Jon Hasselgren, and Tomas Akenine-Möller. High dynamic range texture compression for graphics hardware. *ACM Trans. Graph.*, 25(3):698–706, 2006.

[278] R. Ward N. Sun, H. Mansour. HDR image construction from multi-exposed stereo LDR images. In *17th IEEE International Conference on Image Processing (ICIP)*, pages 2973–2976, September 2010.

[279] Ken-Ichi Naka and William A. H. Rushton. S-potentials from luminosity units in the retina of fish (cyprinidae). *Journal of Physiology*, 185(3):587–599, 1966.

[280] Manish Narwaria, Matthieu Perreira Da Silva, and Patrick Le Callet. Hdr-vqm: An objective quality measure for high dynamic range video. *Signal Processing: Image Communication*, 35:46–60, 2015.

[281] Manish Narwaria, Rafał K. Mantiuk, Mattheiu Perreira Da Silva, and Patrick Le Callet. Hdr-vdp-2.2: a calibrated method for objective quality prediction of high-dynamic range and standard images. *Journal of Electronic Imaging*, 24(1):010501–010501, 2015.

[282] Shree K. Nayar and Vlad Branzoi. Adaptive dynamic range imaging: Optical control of pixel exposures over space and time. In *ICCV '03: Proceedings of the Ninth IEEE International Conference on Computer Vision*, pages 1168–1175, Washington, DC, USA, 2003. IEEE Computer Society.

[283] Shree K. Nayar and Srinivasa G. Narasimhan. Assorted pixels: Multi-sampled imaging with structural models. In *Proceedings of the 7th European Conference on Computer Vision-Part IV*, ECCV '02, pages 636–652, London, UK, UK, 2002. Springer-Verlag.

[284] S.K. Nayar and T. Mitsunaga. High Dynamic Range Imaging: Spatially Varying Pixel Exposures. In *IEEE Conference on Computer Vision and Pattern Recognition (CVPR)*, pages 472–479, Washington, DC, USA, June 2000. IEEE Computer Society.

[285] Laszlo Neumann, Kreimir Matkovic, and Werner Purgathofer. Automatic exposure in computer graphics based on the minimum information loss principle. In *CGI '98: Proceedings of the Computer Graphics International 1998*, page 666, Washington, DC, USA, 1998. IEEE Computer Society.

[286] NextLimits. Maxwell render. Available at http://www.maxwellrender.com/, October 2016.

[287] Ren Ng, Ravi Ramamoorthi, and Pat Hanrahan. All-frequency shadows using non-linear wavelet lighting approximation. *ACM Trans. Graph.*, 22(3):376–381, 2003.

[288] Ren Ng, Ravi Ramamoorthi, and Pat Hanrahan. Triple product wavelet integrals for all-frequency relighting. *ACM Trans. Graph.*, 23(3):477–487, 2004.

[289] J. Nystad, A. Lassen, A. Pomianowski, S. Ellis, and T. Olson. Adaptive scalable texture compression. In *Proceedings of the Fourth ACM SIGGRAPH / Eurographics Conference on High-Performance Graphics*, EGGH-HPG'12, pages 105–114, Aire-la-Ville, Switzerland, 2012. Eurographics Association.

[290] Masahiro Okuda and Nicola Adami. Two-layer coding algorithm for high dynamic range images based on luminance compensation. *J. Vis. Comun. Image Represent.*, 18(5):377–386, 2007.

[291] Omrom. Vision sensors/machine vision systems. Available at https://www.ia.omron.com, October 2016.

[292] opengl.org. OpenGL Image Format. Available at https://www.opengl.org/wiki/Image_Format, October 2016.

[293] Victor Ostromoukhov, Charles Donohue, and Pierre-Marc Jodoin. Fast hierarchical importance sampling with blue noise properties. *ACM Trans. Graph.*, 23(3):488–495, 2004.

[294] Ahmet Oğuz Akyüz and Asli Genctav. A reality check for radiometric camera response recovery algorithms. *Computers & Graphics*, 37(7):935–943, 2013.

[295] Minghao Pan, Rui Wang, Xinguo Liu, Qunsheng Peng, and Hujun Bao. Precomputed radiance transfer field for rendering interreflections in dynamic scenes. *Computer Graphics Forum*, 26(3):485–493, 2007.

[296] Panavision. Genesis digital camera system. Available at http://www.panavision.com/, July 2010.

[297] Sylvain Paris. Edge-preserving smoothing and mean-shift segmentation of video streams. In *Proceedings of the 10th European Conference on Computer Vision: Part II*, ECCV '08, pages 460–473, Berlin, Heidelberg, 2008. Springer-Verlag.

[298] Sylvain Paris, Samuel W. Hasinoff, and Jan Kautz. Local laplacian filters: Edge-aware image processing with a laplacian pyramid. *ACM Trans. Graph.*, 30(4):68:1–68:12, July 2011.

[299] Sung Ho Park and Ethan D. Montag. Evaluating tone mapping algorithms for rendering non-pictorial (scientific) high-dynamic-range images. *J. Vis. Comun. Image Represent.*, 18(5):415–428, 2007.

[300] Steven G. Parker, James Bigler, Andreas Dietrich, Heiko Friedrich, Jared Hoberock, David Luebke, David McAllister, Morgan McGuire, Keith Morley, Austin Robison, and Martin Stich. Optix: A general purpose ray tracing engine. *ACM Trans. Graph.*, 29(4):66:1–66:13, July 2010.

[301] Sumanta N. Pattanaik, James A. Ferwerda, Mark D. Fairchild, and Donald P. Greenberg. A multiscale model of adaptation and spatial vision for

realistic image display. In *SIGGRAPH '98: Proceedings of the 25th Annual Conference on Computer Graphics and Interactive Techniques*, pages 287–298, New York, NY, USA, 1998. ACM.

[302] Sumanta N. Pattanaik, Jack Tumblin, Hector Yee, and Donald P. Greenberg. Time-dependent visual adaptation for fast realistic image display. In *SIGGRAPH '00: Proceedings of the 27th Annual Conference on Computer Graphics and Interactive Techniques*, pages 47–54, New York, NY, USA, 2000. ACM Press/Addison-Wesley Publishing Co.

[303] Sumanta N. Pattanaik and Hector Yee. Adaptive Gain Control for High Dynamic Range Image Display. In *SCCG '02: Proceedings of the 18th Spring Conference on Computer Graphics*, pages 83–87, New York, NY, USA, 2002. ACM.

[304] Fabio Pellacini. envylight: an interface for editing natural illumination. In *ACM SIGGRAPH 2010 papers*, SIGGRAPH '10, pages 34:1–34:8, New York, NY, USA, 2010. ACM.

[305] Patrick Pérez, Michel Gangnet, and Andrew Blake. Poisson image editing. *ACM Trans. Graph.*, 22(3):313–318, 2003.

[306] Ken Perlin and Eric M. Hoffert. Hypertexture. In *Computer Graphics (Proceedings of ACM SIGGRAPH 89)*, pages 253–262, New York, NY, USA, 1989. ACM.

[307] P. Perona and J. Malik. Scale-space and edge detection using anisotropic diffusion. *IEEE Trans. Pattern Anal. Mach. Intell.*, 12(7):629–639, July 1990.

[308] Georg Petschnigg, Richard Szeliski, Maneesh Agrawala, Michael Cohen, Hugues Hoppe, and Kentaro Toyama. Digital photography with flash and no-flash image pairs. *ACM Trans. Graph.*, 23(3):664–672, 2004.

[309] Tuan Q. Pham and Luca J. van Vliet. Separable bilateral filtering for fast video preprocessing. In *IEEE Internat. Conf. on Multimedia & Expo, CD14*, pages 1–4, Los Alamitos, CA, USA, 2005. IEEE Computer Society.

[310] Matt Pharr and Greg Humphreys. *Physically Based Rendering: From Theory to Implementation*. Morgan Kaufmann Publishers Inc., San Francisco, CA, USA, 2004.

[311] Tania Pouli, Alessandro Artusi, Francesco Banterle, Ahmet Oğuz Akyüz, Hans-Peter Seidel, and Erik Reinhard. Color correction for tone reproduction. In *Color and Imaging Conference*. Society for Imaging Science and Technology, November 2013.

[312] William H. Press, Saul A. Teukolsky, William T. Vetterling, and Brian P. Flannery. *Numerical Recipes, Third Edition: The Art of Scientific Computing*. Cambridge University Press, Cambridge, UK, September 2007.

[313] PtGreyResearch. Firefly MV. Available at http://www.ptgrey.com/, October 2016.

[314] Jonathan Ragan-Kelley, Andrew Adams, Sylvain Paris, Marc Levoy, Saman
      Amarasinghe, and Frédo Durand. Decoupling algorithms from schedules
      for easy optimization of image processing pipelines. *ACM Trans. Graph.*,
      31(4):32:1–32:12, July 2012.

[315] Susanto Rahardja, Farzam Farbiz, Corey Manders, Huang Zhiyong, Jamie
      Ng Suat Ling, Ishtiaq Rasool Khan, Ong Ee Ping, and Song Peng. Eye
      hdr: Gaze-adaptive system for displaying high-dynamic-range images. In
      *ACM SIGGRAPH ASIA 2009 Art Gallery & Emerging Technologies:
      Adaptation*, SIGGRAPH ASIA '09, pages 68–68, New York, NY, USA,
      2009. ACM.

[316] Zia ur Rahman, Daniel J. Jobson, and Glenn A. Woodell. Multi-scale
      retinex for color image enhancement. In *Proceedings of the International
      Conference on Image Processing (ICIP)*, pages 1003–1006. IEEE Computer
      Society, 1996.

[317] Ravi Ramamoorthi and Pat Hanrahan. An efficient representation for irra-
      diance environment maps. In *SIGGRAPH '01: Proceedings of the 28th An-
      nual Conference on Computer Graphics and Interactive Techniques*, pages
      497–500, New York, NY, USA, 2001. ACM.

[318] Shanmuganathan Raman and Subhasis Chaudhuri. Bilateral Filter Based
      Compositing for Variable Exposure Photography. In P. Alliez and M. Mag-
      nor, editors, *Eurographics 2009 - Short Papers*. The Eurographics Associ-
      ation, 2009.

[319] Shaun David Ramsey, J. Thomas Johnson, and Charles Hansen. Adaptive
      temporal tone mapping. In *Conference On Computer Graphics and Imaging
      (CGIM)*, pages 143–153, Anaheim, CA, USA, 2004. IASTED.

[320] RedCompany. Red One. Available at http://www.red.com/, October 2016.

[321] Erik Reinhard. Parameter estimation for photographic tone reproduction.
      Journal Graphics Tools, 7(1):45–52, 2002.

[322] Erik Reinhard, Tania Pouli, Timo Kunkel, Ben Long, Anders Ballestad,
      and Gerwin Damberg. Calibrated image appearance reproduction. *ACM
      Trans. Graph.*, 31(6):201:1–201:11, November 2012.

[323] Erik Reinhard, Michael Stark, Peter Shirley, and James Ferwerda. Pho-
      tographic tone reproduction for digital images. *ACM Trans. Graph.*,
      21(3):267–276, 2002.

[324] Allan G. Rempel, Matthew Trentacoste, Helge Seetzen, H. David Young,
      Wolfgang Heidrich, Lorne Whitehead, and Greg Ward. Ldr2hdr: On-the-
      fly reverse tone mapping of legacy video and photographs. *ACM Trans.
      Graph.*, 26(3):39, 2007.

[325] Thomas Richter. Evaluation of floating point image compression. In *Pro-
      ceedings of the 16th IEEE International Conference on Image Processing*,
      ICIP'09, pages 1889–1892, Piscataway, NJ, USA, 2009. IEEE Press.

[326] Thomas Richter, Alessandro Artusi, and Touradj Ebrahimi. JPEG XT: A
      New Family of JPEG Backward-Compatible Standards. *IEEE Multimedia*,
      23:80–88, 2016.

[327] Thomas Richter, Walt Husak, , Niman Ajit, Ten Arkady, Pavel Korshunov, Touradj Ebrahimi, Alessandro Artusi, Massmiliano Agostinelli, Shigetaka Ogawa, Peters Schelkens, Takaaki Ishikawa, and Tim Bruylants. JPEG XT information technology: Scalable compression and coding of continuous-tone still images. http://jpeg.org/jpegxt/index.html, 2015.

[328] Tobias Ritschel and Elmar Eisemann. A computational model of afterimages. *Comput. Graph. Forum*, 31(2):529–534, 2012.

[329] Tobias Ritschel, Matthias Ihrke, Jeppe Revall Frisvad, Joris Coppens, Karol Myszkowski, and Hans-Peter Seidel. Temporal glare: Real-time dynamic simulation of the scattering in the human eye. *Comput. Graph. Forum*, 28(2):183–192, 2009.

[330] Lawrence Roberts. Picture coding using pseudo-random noise. *IEEE Transactions on Information Theory*, 8(2):145–154, February 1962.

[331] Mark A. Robertson, Sean Borman, and Robert L. Stevenson. Dynamic range improvement through multiple exposures. In *Proceedings of the 1999 International Conference on Image Processing (ICIP-99)*, pages 159–163, Los Alamitos, CA, USA, 1999. IEEE Computer Society.

[332] Mark A. Robertson, Sean Borman, and Robert L. Stevenson. Estimation-theoretic approach to dynamic range enhancement using multiple exposures. *Journal of Electronic Imaging*, 12(2):219–228, April 2003.

[333] B. Roch, A. Artusi, D. Michael, Y. Chrysanthou, and A. Chalmers. Interactive local tone mapping operator with the support of graphics hardware. In *Proceedings of the 23rd Spring Conference on Computer Graphics*, SCCG '07, pages 213–218, New York, NY, USA, 2007. ACM.

[334] Kimmo Roimela, Tomi Aarnio, and Joonas Itäranta. High dynamic range texture compression. *ACM Trans. Graph.*, 25(3):707–712, 2006.

[335] Kimmo Roimela, Tomi Aarnio, and Joonas Itäranta. Efficient high dynamic range texture compression. In *SI3D '08: Proceedings of the 2008 Symposium on Interactive 3D Graphics and Games*, pages 207–214, New York, NY, USA, 2008. ACM.

[336] M. Rouf, R. Mantiuk, W. Heidrich, M. Trentacoste, and C. Lau. Glare encoding of high dynamic range images. In *Proceedings of the 2011 IEEE Conference on Computer Vision and Pattern Recognition*, CVPR '11, pages 289–296, Washington, DC, USA, 2011. IEEE Computer Society.

[337] Mushfiqur Rouf, Cheryl Lau, and Wolfgang Heidrich. Gradient domain color restoration of clipped highlights. In *International Workshop on Projector-Camera Systems (PROCAMS)*. IEEE Computer Society, 2012.

[338] Alexa I. Ruppertsberg, Marina Bloj, Francesco Banterle, and Alan Chalmers. Displaying colourimetrically calibrated images on a high dynamic range display. *Journal Visual Communication Image Represent.*, 18(5):429–438, 2007.

[339] Imari Sato, Yoichi Sato, and Katsushi Ikeuchi. Acquiring a radiance distribution to superimpose virtual objects onto a real scene. *IEEE Transactions on Visualization and Computer Graphics*, 5(1):1–12, 1999.

[340] A. Scheel, M. Stamminger, and H.-P. Seidel. Tone Reproduction for Interactive Walkthroughs. *Computer Graphics Forum*, 19(3):301–311, 2000.

[341] Christophe Schlick. Quantization techniques for visualization of high dynamic range pictures. In *Proceeding of the Fifth Eurographics Workshop on Rendering*, pages 7–18, June 1994.

[342] Helge Seetzen, Wolfgang Heidrich, Wolfgang Stuerzlinger, Greg Ward, Lorne Whitehead, Matthew Trentacoste, Abhijeet Ghosh, and Andrejs Vorozcovs. High dynamic range display systems. *ACM Trans. Graph.*, 23(3):760–768, 2004.

[343] Pradeep Sen, Nima Khademi Kalantari, Maziar Yaesoubi, Soheil Darabi, Dan B. Goldman, and Eli Shechtman. Robust patch-based HDR reconstruction of dynamic scenes. *ACM Transactions on Graphics (Proceedings of SIGGRAPH Asia 2012)*, 31(6):203:1–203:11, November 2012.

[344] Ana Serrano, Felix Heide, Diego Gutierrez, Gordon Wetzstein, and Belen Masia. Convolutional sparse coding for high dynamic range imaging. *Computer Graphics Forum*, 35(2):153–163, 2016.

[345] Qi Shan, Tony DeRose, and John Anderson. Tone mapping high dynamic range videos using wavelets. Technical report, Pixar Ltd, 2012.

[346] Shizhe Shen. *Color Difference Formula and Uniform Color Space Modeling and Evaluation*. PhD thesis, Rochester Institute of Technology, 2009.

[347] Peter-Pike Sloan, Jan Kautz, and John Snyder. Precomputed radiance transfer for real-time rendering in dynamic, low-frequency lighting environments. *ACM Trans. Graph.*, 21(3):527–536, 2002.

[348] Kaleigh Smith, Grzegorz Krawczyk, Karol Myszkowski, and Hans-Peter Seidel. Beyond tone mapping: Enhanced depiction of tone mapped HDR images. *Computer Graphics Forum (Proceeding of Eurographics 2006)*, 25(3):427–438, September 2006.

[349] Stephen M. Smith and J. Michael Brady. SUSAN: A New Approach to Low Level Image Processing. *Int. J. Comput. Vision*, 23(1):45–78, May 1997.

[350] Sony. F65. Available at https://pro.sony.com/bbsc/ssr/product-F65RSPAC1/, October 2016.

[351] Spheron. Spheron HDR VR. Available at http://www.spheron.com/, December 2008.

[352] Sam Staton, Kurt Debattista, Thomas Bashford-Rogers, and Alan Chalmers. Light clustering for dynamic image based lighting. In *Theory and Practice of Computer Graphics, Rutherford, United Kingdom, 2012. Proceedings*, pages 17–24, 2012.

[353] J.C. Stevens and Stanley S. Stevens. Brightness function: Effects of adaptation. *Journal Optical Society of America*, 53(3):375–385, March 1963.

[354] Stanley S. Stevens and J.C. Stevens. Brightness function: Parametric effects of adaptation and contrast. *Journal Optical Society of America*, 50(11):1139, November 1960.

[355] Michael Stokes, Matthew Anderson, Srinivasan Chandrasekar, and Ricardo Motta. A Standard Default Color Space for the Internet—sRGB. Available at http://www.w3.org/Graphics/Color/sRGB.html, November 1996.

[356] Eric J. Stollnitz, Tony D. DeRose, and David H. Salesin. Wavelets for computer graphics: A primer. *IEEE Comput. Graph. Appl.*, 15(3):76–84, 1995.

[357] Kartic Subr, Cyril Soler, and Frédo Durand. Edge-preserving multiscale image decomposition based on local extrema. *ACM Trans. Graph.*, 28(5):147:1–147:9, December 2009.

[358] Deqing Sun, Stefan Roth, and Michael J. Black. Secrets of optical flow estimation and their principles. In *IEEE Conf. on Computer Vision and Pattern Recognition (CVPR)*, pages 2432–2439. IEEE, June 2010.

[359] Wen Sun, Yan Lu, Feng Wu, and Shipeng Li. Dhtc: An effective dxtc-based hdr texture compression scheme. In *GH '08: Proceedings of the 23rd ACM Siggraph/Eurographics Symposium on Graphics Hardware*, pages 85–94, Aire-la-Ville, Switzerland, 2008. Eurographics Association.

[360] Wen Sun, Yan Lu, Feng Wu, Shipeng Li, and John Tardif. High-dynamic-range texture compression for rendering systems of different capacities. *IEEE Trans. Vis. Comput. Graph.*, 16(1):57–69, 2010.

[361] Richard Szeliski. *Computer Vision: Algorithms and Applications*. Springer-Verlag, New York, NY, USA, 2010.

[362] Hiroyuki Takeda, Sina Farsiu, and Peyman Milanfar. Higher order bilateral filters and their properties. In Charles A. Bouman, Eric L. Miller, and Ilya Pollak, editors, *Computational Imaging V, January 29–31*, volume 6498 of *SPIE Proceedings*, page 64980S. SPIE, 2007.

[363] Justin Talbot, David Cline, and Parris K. Egbert. Importance resampling for global illumination. In *Rendering Techniques 2005 Eurographics Symposium on Rendering*, pages 139–146, Aire-la-Ville, Switzerland, 2005. Eurographics Association.

[364] Eino-Ville Talvala, Andrew Adams, Mark Horowitz, and Marc Levoy. Veiling glare in high dynamic range imaging. *ACM Trans. Graph.*, 26(3), July 2007.

[365] Chris Tchou, Jessi Stumpfel, Per Einarsson, Marcos Fajardo, and Paul Debevec. Unlighting the parthenon. In *SIGGRAPH '04: ACM Siggraph 2004 Sketches*, page 80, New York, NY, USA, 2004. ACM.

[366] William B. Thompson, Peter Shirley, and James A. Ferwerda. A spatial post-processing algorithm for images of night scenes. *J. Graphics, GPU, & Game Tools*, 7(1):1–12, 2002.

[367] Michael D. Tocci, Chris Kiser, Nora Tocci, and Pradeep Sen. A versatile HDR video production system. *ACM Transactions on Graphics*, 30(4):41:1–41:10, July 2011.

[368] Carlo Tomasi and Roberto Manduchi. Bilateral filtering for gray and color images. In *ICCV '98: Proceedings of the Sixth International Conference on Computer Vision*, page 839, Washington, DC, USA, 1998. IEEE Computer Society.

[369] Matthew Trentacoste, Wolfgang Heidrich, Lorne Whitehead, Helge Seetzen, and Greg Ward. Photometric image processing for high dynamic range displays. *Journal Visual Communication Image Represent.*, 18(5):439–451, 2007.

[370] Chengjie Tu Tu, Sridhar Srinivasan, Gary J. Sullivan, Shankar Regunathan, and Rico Malvar. Low-complexity hierarchical lapped transform for lossy-to-lossless image coding in jpeg xr / hd photo. In *Proc. SPIE 7073, Applications of Digital Image Processing XXXI, 70730C*, pages 70730C–1–70730C–12. SPIE, August 2008.

[371] Jack Tumblin, Amit Agrawal, and Ramesh Raskar. Why i want a new. In *Proceedings of the 2005 IEEE Computer Society Conference on Computer Vision and Pattern Recognition (CVPR'05) - Volume 1 - Volume 01*, CVPR '05, pages 103–110, Washington, DC, USA, 2005. IEEE Computer Society.

[372] Jack Tumblin, Jessica K. Hodgins, and Brian K. Guenter. Two methods for display of high contrast images. *ACM Trans. Graph.*, 18(1):56–94, 1999.

[373] Jack Tumblin and Holly Rushmeier. Tone reproduction for realistic images. *IEEE Comput. Graph. Appl.*, 13(6):42–48, 1993.

[374] Jack Tumblin and Greg Turk. Lcis: A boundary hierarchy for detail-preserving contrast reduction. In *SIGGRAPH '99: Proceedings of the 26th Annual Conference on Computer Graphics and Interactive Techniques*, pages 83–90, New York, NY, USA, 1999. ACM Press/Addison-Wesley Publishing Co.

[375] Okan Tarhan Tursun, Ahmet Oğuz Akyüz, Aykut Erdem, and Erkut Erdem. The state of the art in HDR deghosting: A survey and evaluation. *Computer Graphics Forum (Proc. of Eurographics 2015)*, 34(2):683–707, May 2015.

[376] Jonas Unger and Stefan Gustavson. High dynamic range video for photometric measurement of illumination. In *Proceedings of Sensors, Cameras and Systems for Scientific/Industrial Applications X, IS&T/SPIE 19th Inernational Symposium on Electronic Imaging*. SPIE, 2007.

[377] Jonas Unger, Stefan Gustavson, Per Larsson, and Anders Ynnerman. Free form incident light fields. *Computer Graphics Forum (Proc. of Eurographics Symposium on Rendering 2008)*, 27(4):1293–1301, June 2008.

[378] Jonas Unger, Joel Kronander, Per Larsson, Stefan Gustavson, Joakim Löw, and Anders Ynnerman. Spatially Varying Image Based Lighting using HDR-video. *Computers and Graphics*, 37(7), 2013.

[379] Jonas Unger, Joel Kronander, Per Larsson, Stefan Gustavson, and Anders Ynnerman. Temporally and Spatially Varying Image Based Lighting using HDR-video. In *EUSIPCO*, 2013.

[380] Jonas Unger, Anders Wenger, Tim Hawkins, A. Gardner, and Paul Debevec. Capturing and rendering with incident light fields. In *EGRW '03: Proceedings of the 14th Eurographics Workshop on Rendering*, pages 141–149, Aire-la-Ville, Switzerland, 2003. Eurographics Association.

[381] International Telecommunication Union. Recommendation ITU-R BT. 601 - 7, 2011.

[382] International Telecommunication Union. Recommendation ITU-R BT. 2020 - 2, 2015.

[383] International Telecommunication Union. Recommendation ITU-R BT. 709 - 6, 2015.

[384] International Telecommunication Union. Recommendation ITU-R BT. 2100, 2016.

[385] C Urbano, Luís Magalhães, J Moura, Massimo Bessa, A Marcos, and Alan Chalmers. Tone mapping operators on small screen devices: an evaluation study. *Computer Graphics Forum*, 29(8):2469–2478, 2010.

[386] T. Nguyen V. Ramachandra, M. Zwicker. HDR imaging from differently exposed multiview videos. In *3DTV Conference: The True Vision-Capture, Transmission and Display of 3D Video*, May 2008.

[387] J. Hans Van Hateren. Encoding of high dynamic range video with a model of human cones. *ACM Trans. Graph.*, 25(4):1380–1399, 2006.

[388] Peter Vangorp, Karol Myszkowski, Erich W. Graf, and Rafał K. Mantiuk. A model of local adaptation. *ACM Trans. Graph.*, 34(6):166:1–166:13, October 2015.

[389] Vladimir N. Vapnik. *The Nature of Statistical Learning Theory*. Springer-Verlag, New York, NY, USA, 1995.

[390] D. Varghese, R. Wanat, and R.K. Mantiuk. Colorimetric calibration of high dynamic range images with a colorchecker chart. In *HDRi 2014: Second International Conference and SME Workshop on HDR Imaging*, pages 17–22, 2014.

[391] Eric Veach and Leonidas J. Guibas. Optimally combining sampling techniques for monte carlo rendering. In *SIGGRAPH '95: Proceedings of the 22nd Annual Conference on Computer Graphics and Interactive Techniques*, pages 419–428, New York, NY, USA, 1995. ACM.

[392] Eric Veach and Leonidas J. Guibas. Metropolis light transport. In *Proceedings of the 24th Annual Conference on Computer Graphics and Interactive Techniques*, SIGGRAPH '97, pages 65–76, New York, NY, USA, 1997. ACM Press/Addison-Wesley Publishing Co.

[393] Kuntee Viriyothai and Paul Debevec. Variance minimization light probe sampling. In *SIGGRAPH '09: Posters*, SIGGRAPH '09, pages 92:1–92:1, New York, NY, USA, 2009. ACM.

[394] Silicon Vision. Silicon vision lars iii. Available at http://www.si-vision. com/, May 2010.

[395] VisionResearch. Phantom HD. Available at http://www.visionresearch.
com/, October 2016.

[396] Elena Šikudova, Tania Pouli, Alessandro Artusi, Akyüz Ahmet Oğuz, Banterle Francesco, Erik Reinhard, and Zeynep Miray Mazlumoglu. A Gamut
Mapping Framework for Color-Accurate Reproduction of HDR Images.
IEEE Transaction on Computer Graphics and Applications, 36(4):78–90,
2016.

[397] Jan Walraven and J. Mathé Valeton. Visual adaptation and response saturation. In W. A. Van de Grind and J. J. Koenderink, editors, *Limits in
Perception*, pages 401–429, The Netherlands, 1984. VNU Science Press.

[398] Bruce Walter, Sebastian Fernandez, Adam Arbree, Kavita Bala, Michael
Donikian, and Donald P. Greenberg. Lightcuts: A scalable approach to
illumination. *ACM Trans. Graph.*, 24(3):1098–1107, 2005.

[399] Liang Wan, Tien-Tsin Wong, and Chi-Sing Leung. Spherical q2-tree for
sampling dynamic environment sequences. In *Proceedings of the Sixteenth
Eurographics Conference on Rendering Techniques*, EGSR '05, pages 21–30,
Aire-la-Ville, Switzerland, 2005. Eurographics Association.

[400] Robert Wanat and Rafał K. Mantiuk. Simulating and compensating
changes in appearance between day and night vision. *ACM Trans. Graph.*,
33(4):147:1–147:12, July 2014.

[401] Lvdi Wang, Xi Wang, Peter-Pike Sloan, Li-Yi Wei, Xin Tong, and Baining
Guo. Rendering from compressed high dynamic range textures on programmable graphics hardware. In *I3D '07: Proceedings of the 2007 Symposium on Interactive 3D Graphics and Games*, pages 17–24, New York,
NY, USA, 2007. ACM.

[402] Lvdi Wang, Li-Yi Wei, Kun Zhou, Baining Guo, and Heung-Yeung Shum.
High dynamic range image hallucination. In *SIGGRAPH '07: ACM SIG-
GRAPH 2007 Sketches*, page 72, New York, NY, USA, 2007. ACM.

[403] Zhou Wang and Alan Bovik. A universal image quality index. *IEEE Signal
Processing Letters*, 9(3):81–84, March 2002.

[404] Zhou Wang, Alan C Bovik, Hamid R Sheikh, and Eero P Simoncelli. Image
quality assessment: from error visibility to structural similarity. *IEEE
Transactions on Image Processing*, 13(4):600–612, 2004.

[405] Greg Ward. Real pixels. *Graphics Gems*, 2:15–31, 1991.

[406] Greg Ward. A contrast-based scalefactor for luminance display. In Paul
Heckbert, editor, *Graphics Gems IV*, chapter VII.2, pages 415–421. Academic Press, Boston, MA, USA, 1994.

[407] Greg Ward. The radiance lighting simulation and rendering system. In
*SIGGRAPH '94: Proceedings of the 21st Annual Conference on Computer
Graphics and Interactive Techniques*, pages 459–472, New York, NY, USA,
1994. ACM.

[408] Greg Ward. A wide field, high dynamic range, stereographic viewer. In
Proceeding of PICS 2002, 2002.

[409] Greg Ward and Maryann Simmons. Subband encoding of high dynamic range imagery. In *APGV '04: Proceedings of the 1st Symposium on Applied Perception in Graphics and Visualization*, pages 83–90, New York, NY, USA, 2004. ACM Press.

[410] Greg Ward and Maryann Simmons. JPEG-HDR: A backwards-compatible, high dynamic range extension to JPEG. In *In Proceedings of the 13th Color Imaging Conference*, pages 283–290. Society for Imaging Science and Technology, 2005.

[411] Andrew B. Watson. Temporal sensitivity. In Kenneth R. Boff, Lloyd Kaufman, and James P. Thomas, editors, *Handbook of Perception and Human Performance, Volume I*, chapter 6, pages 1–43. John Wiley & Sons, New York, NY, USA, 1986.

[412] Andrew B. Watson. The cortex transform: Rapid computation of simulated neural images. *Comput. Vision Graph. Image Process.*, 39(3):311–327, 1987.

[413] Andrew B. Watson and Joshua A. Solomon. Model of visual contrast gain control and pattern masking. *Journal of the Optical Society of America*, 14(9):2379–2391, 1997.

[414] Ben Weiss. Fast median and bilateral filtering. *ACM Trans. Graph.*, 25(3):519–526, 2006.

[415] Tomihisa Welsh, Michael Ashikhmin, and Klaus Mueller. Transferring color to greyscale images. *ACM Trans. Graph.*, 21(3):277–280, 2002.

[416] Gordon Wetzstein, Douglas Lanman, Wolfgang Heidrich, and Ramesh Raskar. Layered 3d: Tomographic image synthesis for attenuation-based light field and high dynamic range displays. *ACM Trans. Graph.*, 30(4):95:1–95:12, July 2011.

[417] Turner Whitted. An improved illumination model for shaded display. In *SIGGRAPH '79: Proceedings of the 6th Annual Conference on Computer Graphics and Interactive Techniques*, page 14, New York, NY, USA, 1979. ACM.

[418] Thomas Wiegand, Gary J. Sullivan, Gisle Bjontegaard, and Ajay Luthra. Overview of the H.264/AVC video coding standard. *IEEE Transactions on Circuits and Systems for Video Technology*, 13(7):560–576, July 2003.

[419] Bennett Wilburn, Neel Joshi, Vaibhav Vaish, Eino-Ville Talvala, Emilio Antunez, Adam Barth, Andrew Adams, Mark Horowitz, and Marc Levoy. High performance imaging using large camera arrays. *ACM Trans. Graph.*, 24(3):765–776, July 2005.

[420] Lance Williams. Casting curved shadows on curved surfaces. In *SIGGRAPH '78: Proceedings of the 5th Annual Conference on Computer Graphics and Interactive Techniques*, pages 270–274, New York, NY, USA, 1978. ACM.

[421] Hugh R. Wilson. Psychophysical models of spatial vision and hyperacuity. In D. Regan, editor, *Spatial Vision*, pages 64–81. CRC Press, Boca Raton, FL, USA, 1991.

[422] Günter Wyszecki and Walter Stanley Stiles. *Color Science: Concepts and Methods, Quantitative Data and Formulae*. Wiley, New York, NY, USA, 2000.

[423] Di Xu, Colin Doutre, and Panos Nasiopoulos. Correction of clipped pixels in color images. *IEEE Transactions on Visualization and Computer Graphics*, 17(3):333–344, May 2011.

[424] Li Xu, Cewu Lu, Yi Xu, and Jiaya Jia. Image smoothing via l0 gradient minimization. *ACM Trans. Graph.*, 30(6):174:1–174:12, December 2011.

[425] Ruifeng Xu, Sumanta N. Pattanaik, and Charles E. Hughes. High-dynamic-range still-image encoding in jpeg 2000. *IEEE Computer Graphics and Applications*, 25(6):57–64, 2005.

[426] Xvid. Xvid. Available at http://www.xvid.com/, February 2010.

[427] Xuan Yang, Linling Zhang, Tien-Tsin Wong, and Pheng-Ann Heng. Binocular tone mapping. *ACM Trans. Graph.*, 31(4):93:1–93:10, July 2012.

[428] Hector Yee and Sumanta N. Pattanaik. Segmentation and adaptive assimilation for detail-preserving display of high-dynamic range images. *The Visual Computer*, 19(7/8):457–466, December 2003.

[429] Hojatollah Yeganeh and Zhou Wang. Objective quality assessment of tone-mapped images. *IEEE Transactions on Image Processing*, 22(2):657–667, 2013.

[430] Akiko Yoshida, Volker Blanz, Karol Myszkowski, and Hans-Peter Seidel. Perceptual evaluation of tone mapping operators with real-world scenes. In Bernice E. Rogowitz, Thrasyvoulos N. Pappas, and Scott J. Daly, editors, *Human Vision and Electronic Imaging X, IS&T/SPIE's 17th Annual Symposium on Electronic Imaging*, pages 192–203. SPIE, January 2005.

[431] Akiko Yoshida, Volker Blanz, Karol Myszkowski, and Hans-Peter Seidel. Testing tone mapping operators with human-perceived reality. *Journal of Electronic Imaging*, 16(1):013004–1–013004–14, January–March 2007.

[432] Akiko Yoshida, Rafał Mantiuk, Karol Myszkowski, and Hans-Peter Seidel. Analysis of reproducing real-world appearance on displays of varying dynamic range. *Computer Graphics Forum*, 25(3):415–426, September 2006.

[433] Yonghao Yue, Kei Iwasaki, Bing-Yu Chen, Yoshinori Dobashi, and and-Tomoyuki Nishita. Interactive rendering of interior scenes with dynamic environment illumination. *Computer Graphics Forum*, 28(7):1935–1944, October 2009.

[434] Wei Zhang and Wai-Kuen Cham. Reference-guided exposure fusion in dynamic scenes. *J. Vis. Comun. Image Represent.*, 23(3):467–475, April 2012.

[435] Hang Zhao, Boxin Shi, Christy Fernandez-Cull, Sai-Kit Yeung, and Ramesh Raskar. Unbounded high dynamic range photography using a modulo camera. In *Proceedings of the International Conference on Computational Photography (ICCP)*, pages 1–10. IEEE Computer Society, 2015.

[436] Kun Zhou, Yaohua Hu, Stephen Lin, Baining Guo, and Heung-Yeung Shum. Precomputed shadow fields for dynamic scenes. *ACM Trans. Graph.*, 24(3):1196–1201, 2005.

Index